HARDROCK MINING ON FEDERAL LANDS

Committee on Hardrock Mining on Federal Lands

Committee on Earth Resources

Board on Earth Sciences and Resources

Commission on Geosciences, Environment, and Resources

National Research Council

NATIONAL ACADEMY PRESS
Washington, D.C.

NOTICE: The project that is the subject of this report was approved by the Governing Board of the National Research Council, whose members are drawn from the councils of the National Academy of Sciences, the National Academy of Engineering, and the Institute of Medicine. The members of the committee responsible for the report were chosen for their special competences and with regard for appropriate balance.

This study was supported by the Bureau of Land Management, Department of the Interior, under assistance award No. 1434-HQ-97-AG-01886. The views and conclusions contained in this document are those of the authors and should not be interpreted as necessarily representing the official policies, either expressed or implied, of the U.S. government.

International Standard Book Number 0-309-06596-8

Additional copies of this report are available from:

National Academy Press
2101 Constitution Avenue, N.W.
Box 285
Washington, DC 20055
800-624-6242
202-334-3313 (in the Washington metropolitan area)
http://www.nap.edu

Cover: Photograph of Gold Quarry Mine, Newmont Mining Company, Carlin, NV courtesy of Raymond Krauss.

Printed in the United States of America

COMMITTEE ON HARDROCK MINING ON FEDERAL LANDS

COMMITTEE ON EARTH RESOURCES

This report has been reviewed by individuals chosen for their diverse perspectives and technical expertise in accordance with procedures approved by the NRC's Report Review Committee. The purpose of this independent review is to provide candid and critical comments that will assist the authors and the NRC in making their published report as sound as possible and to ensure that the report meets institutional standards for objectivity, evidence, and responsiveness to the study charge. The content of the review comments and draft manuscript remain confidential to protect the integrity of the deliberative process. We wish to thank the following individuals for their participation in the review of this report:

Edie B. Allen
University of California,
  Riverside

R. Ray Beebe
Consultant
Tucson, Arizona

John S. Chipman
University of Minnesota
Minneapolis

David H. Getches
University of Colorado School of
  Law, Boulder

George M. Hornberger
University of Virginia
Charlottesville

J.S. Livermore
Cordilleran Exploration
  Company
Reno, Nevada

Glenn C. Miller
University of Nevada, Reno

Dianne R. Nielson
Utah Department of
  Environmental Quality
Salt Lake City

Raymond A. Price
Queen's University
Kingston, Ontario

Steven P. Quarles
Crowell & Moring LLP
Washington, D.C.

Although the individuals listed above have provided many constructive comments and suggestions, responsibility for the final content of this report rests solely with the authoring committee and the NRC.

The National Academy of Sciences is a private, nonprofit, self-perpetuating society of distinguished scholars engaged in scientific and engineering research, dedicated to the furtherance of science and technology and to their use for the general welfare. Upon the authority of the charter granted to it by the Congress in 1863, the Academy has a mandate that requires it to advise the federal government on scientific and technical matters. Dr. Bruce Alberts is president of the National Academy of Sciences.

The National Academy of Engineering was established in 1964, under the charter of the National Academy of Sciences, as a parallel organization of outstanding engineers. It is autonomous in its administration and in the selection of its members, sharing with the National Academy of Sciences the responsibility for advising the federal government. The National Academy of Engineering also sponsors engineering programs aimed at meeting national needs, encourages education and research, and recognizes the superior achievements of engineers. Dr. William A. Wulf is president of the National Academy of Engineering.

The Institute of Medicine was established in 1970 by the National Academy of Sciences to secure the services of eminent members of appropriate professions in the examination of policy matters pertaining to the health of the public. The Institute acts under the responsibility given to the National Academy of Sciences by its congressional charter to be an adviser to the federal government and, upon its own initiative, to identify issues of medical care, research, and education. Dr. Kenneth I. Shine is president of the Institute of Medicine.

The National Research Council was organized by the National Academy of Sciences in 1916 to associate the broad community of science and technology with the Academy's purposes of furthering knowledge and advising the federal government. Functioning in accordance with general policies determined by the Academy, the Council has become the principal operating agency of both the National Academy of Sciences and the National Academy of Engineering in providing services to the government, the public, and the scientific and engineering communities. The Council is administered jointly by both Academies and the Institute of Medicine. Dr. Bruce Alberts and Dr. William A. Wulf are chairman and vice-chairman, respectively, of the National Research Council.

Contents

Executive Summary

This report responds to a request by Congress that the National Research Council assess the adequacy of the regulatory framework for hardrock mining on federal lands. The regulatory framework applies to hardrock (locatable) minerals—such as gold, silver, copper, and uranium—on over 350 million acres of federal lands in the western United States. To conduct the study, the National Research Council appointed the Committee on Hardrock Mining on Federal Lands in January 1999.

The charge to the Committee had three major components. First, the Committee was asked to identify federal and state statutes and regulations applicable to environmental protection of federal lands in connection with mining activities. Second, the Committee was charged with considering the adequacy of statutes and regulations to prevent unnecessary or undue degradation of the federal lands. Third, the Committee was asked for its recommendations for the coordination of federal and state regulations to ensure environmental protection, increase efficiency, avoid duplication and delay, and identify the most cost-effective manner for implementation.

The report deals with hardrock mining on federal lands managed by two agencies—the Bureau of Land Management (BLM) in the Department of the Interior, and the Forest Service in the Department of Agriculture. The area of federal land available to hardrock mining in the western states is enormous, but the surface area actually physically disturbed by active mining is small in comparison. However, impacts on water quality, vegetation, and aquatic biota often extend beyond the immediate area of the mine site. The BLM is responsible for 260 million acres of land in the western states, including Alaska, of which roughly 90% are open to hardrock mining. Approximately 0.06% of BLM lands are affected by active mining and mineral exploration operations. The Forest Service manages 163 million acres in the western states, of which roughly 80% are open to hardrock mining. Together, the two land management agencies are responsible for 38% of the total area of the western states. These lands are important for their potential mineral wealth and

1

timber, for grazing purposes, as a source of clean water, as a location for recreational activities, as wildlife habitats and scenic areas, and for other purposes.

Mining inevitably affects these resources. The significance of potential negative impacts depends on the extent to which they can be avoided or mitigated. This, to some extent, depends on compliance with regulations. Mining, its impacts on other resources and uses, and the regulatory structure are related matters that require balance and reason when dealing with the potentially competing issues of protection of the environment, production of minerals and metals and employment for society, and associated federal and state statutory responsibilities.

This report is particularly timely in that the BLM has proposed to revise its regulations for hardrock mineral exploration and development conducted under the authority of the General Mining Law of 1872. Because the Committee's charge stated that the baseline for the study was the existing regulatory framework rather than proposed changes to that framework, the Committee did not focus on the BLM proposal to revise its regulations. The Committee was told that the Forest Service was also internally considering revisions in its regulations for hardrock mining in national forests, but the Committee has not reviewed any of the changes being considered.

The Committee's conclusions and recommendations are based on information obtained in a series of presentations and open public forums at Committee meetings in Washington, D.C., Denver, Reno, and Spokane. Committee members also made visits to a variety of mining operations in Colorado, California, Nevada, Washington, and Alaska. The Committee also was helped in its work by a large volume of reports, submissions to the Committee, and copies of submissions to the BLM in connection with the proposed revision of its hardrock mining regulations. The Committee itself represented a wide spectrum of skills and experience relevant to mining on federal lands.

HARDROCK MINING AND THE ENVIRONMENT

Hardrock mining occurs where minerals are concentrated in economically viable deposits. Ore deposits form as variants of such geologic processes as volcanism, weathering, and sedimentation operating with an extraordinary intensity. Ore deposits typically are parts of large-scale (several miles across and perhaps just as deep) ore-forming systems in which many elements, not just those of economic interest, have been enriched. Only a very small portion of Earth's continental crust (less than 0.01%) contains economically viable mineral deposits. Thus, mines can only be located in those few places where economically viable deposits were formed and discovered.

Many hardrock commodities are associated with magmatic and hydrothermal processes, which in turn, are associated with modern or ancient mountain belts. The abundant igneous rocks and associated hydrothermal systems and the mountainous or sparsely vegetated terrain make the West the location of most hardrock mines in the United States. Some of these same areas are also valued for aesthetic and cultural reasons, which creates potential for conflict among uses of the land. While society requires a healthy environment, it also requires sources of materials, many of which can be supplied only by mining.

The mining process consists of exploration, mine development, mining (extraction), mineral processing (beneficiation), and reclamation (including post-closure). The hardrock mining process is described in Chapter 1 and Appendix A. Each step from exploration through post-closure has the potential to cause environment impacts. In addition to the obvious disturbance of the land surface, mining may affect, to varying degrees, groundwater, surface water, aquatic biota, aquatic and terrestrial vegetation, wildlife, soils, air, and cultural resources. Actions based on environmental regulations may avoid, limit, control, or offset many of these potential impacts, but mining will, to some degree, always alter landscapes and environmental resources. Regulations intended to control and manage these alterations of the landscape and the environment in an acceptable way are generally in place and are updated as new technologies are developed to improve mineral extraction, to reclaim mined lands, and to limit environmental impacts. Therefore, the committee emphasizes that these potential impacts will not necessarily occur, and when they do, they will not occur with the same intensity in all cases. The potential environmental impacts of hardrock mining are discussed in Chapter 1 and Appendix B.

EXISTING REGULATORY FRAMEWORK

Hardrock mining operations in the United States are regulated by a complex set of federal and state laws and regulations intended to protect the environment (see Chapter 2). The scope and degree of regulation depends on the type and size of the mining operation; the kinds of land, water, and biological resources affected; the state in which the operation is located; the organization of the state and local permitting agencies; and the ways federal and state agencies implement relevant statutes and regulations. The basic statute for hardrock mining on federal lands is the General Mining Law of 1872. Land management direction is provided in the Federal Land Policy and

Management Act of 1976 (FLPMA) for the BLM and in the 1897 Organic Act and the 1976 National Forest Management Act for the Forest Service. These statutory authorities find further expression in the regulations adopted by the respective agencies.

Proposed mining activities on federal lands trigger the application of BLM's Part 3809 regulations (43 CFR Part 3809) and the Forest Service's Part 228 regulations (36 CFR Part 228). BLM's Part 3809 regulations establish guidelines intended to assure compliance with the FLPMA prohibition of "unnecessary or undue degradation of public lands." The Forest Service's Part 228 regulations establish guidelines intended to assure compliance with the Forest Service regulatory requirement to "minimize adverse environmental impacts on national forest surface resources," based on the Organic Act.

The National Environmental Policy Act (NEPA) serves to integrate BLM and Forest Service decision making on particular mining proposals with evaluation of other environmental concerns, as well as with other state and federal permitting requirements (see Chapter 2). For operations on federal lands that are expected to have significant impacts on the environment, the environmental impact statement (EIS) under NEPA serves as the "spine" of the federal land manager's decision-making process. The EIS process includes requirements for publicly "scoping" the issues and identifying alternatives to be evaluated, and results in a record of decision that determines the content of the plan of operations and mitigation requirements. For smaller operations on federal lands, an environmental assessment (EA) often is produced instead of an EIS. The EA is intended to assist the federal land management agency in deciding whether environmental impacts will be significant.

Various state and federal laws establish environmental requirements applicable to mining operations on federal lands. These include state reclamation laws, state and federal water pollution laws, state groundwater quality laws, state water rights laws, state and federal fish and wildlife laws, state and federal air quality laws, wetlands laws, and other laws. Memoranda of understanding (MOUs) among the federal agencies and state agencies establish the links between state environmental requirements and federal land managers' decisions.

Conclusions and recommendations are presented below related to the coordination of federal and state regulations to ensure environmental protection, increase efficiency, avoid duplication and delay, and identify the most cost-effective manner for implementation. Each conclusion is followed by a recommendation. Conclusions and recommendations are discussed in more detail, including justifications and discussion, in Chapter 4.

GENERAL CONCLUSIONS AND RECOMMENDATIONS

Existing regulations are generally well coordinated, although some changes are necessary. The overall structure of the federal and state laws and regulations that provide mining-related environmental protection is complicated but generally effective. The structure reflects regulatory responses to geographical differences in mineral distribution among the states, as well as the diversity of site-specific environmental conditions. It also reflects the unique and overlapping federal and state responsibilities.

Conclusion: Federal land management agencies' regulatory standards for mining should continue to focus on the clear statement of management goals rather than on defining inflexible, technically prescriptive standards. Simple "one-size-fits-all" solutions are impractical because mining confronts too great an assortment of site-specific technical, environmental, and social conditions. Each proposed mining operation should be examined on its own merits. For example, if backfilling of mines is to be considered, it should be determined on a case-by-case basis, as was concluded by the Committee on Surface Mining and Reclamation (COSMAR) report (NRC, 1979). **Recommendation: BLM and the Forest Service should continue to base their permitting decisions on the site-specific evaluation process provided by NEPA. The two land management agencies should continue to use comprehensive performance-based standards rather than using rigid, technically prescriptive standards. The agencies should regularly update technical and policy guidance documents to clarify how statutes and regulations should be interpreted and enforced.**

Although the overall regulatory structure for hardrock mining on federal lands is effective, the Committee has identified a number of areas where the implementation of existing laws and regulations could be improved (see "Improvements in Implementation" below). The Committee also has identified specific issues or "gaps" in the existing regulations intended to protect the environment (see "Regulatory Gaps" below). In addition, the Committee recommends that research be conducted to improve scientific understanding of issues related to environmental impacts of hardrock mining (see "Research Needs" below).

IMPROVEMENTS IN IMPLEMENTATION

Improvements in the implementation of existing regulations present the greatest opportunity for improving environmental protection and the efficiency of the regulatory process. Federal land management agencies already have at their disposal an array of statutes and regulations that for the most part assure environmentally responsible resource development, but these tools are unevenly and sometimes inexpertly applied. The committee has identified the following issues where the implementation of existing regulations could be improved through enhanced information management, greater understanding of existing laws and regulations by federal land management staff, and improved efficiency:

Conclusion: The Committee was consistently frustrated by the lack of reliable information on mining on federal lands. The lack of thorough information extends from that needed to characterize the lands available for mineral development to that needed to track mining and compliance with regulations. Without more and better information, it is difficult to manage federal lands properly and assure the public that its interests are protected. **Recommendation: BLM and the Forest Service should maintain a management information system that effectively tracks compliance with operating plans and environmental permits, and communicates this information to agency managers, the interested public, and other stakeholders.**

Conclusion: Better information on federal lands is needed to make wise land use decisions. The land use planning process required for BLM and Forest Service lands by the Federal Lands Policy and Management Act and the National Forest Management Act, respectively, provide for identification of land and resources deserving special environmental concern. **Recommendation: BLM and the Forest Service should identify, regularly update, and make available to the public, information identifying those parts of federal lands that will require special consideration in land use decisions because of natural and cultural resources or special environmental sensitivities.**

Conclusion: The NEPA process is the key to establishing an effective balance between mineral development and environmental protection. The effectiveness of NEPA depends on the full participation of all stakeholders throughout the NEPA process. Unfortunately this rarely happens in a timely fashion. **Recommendation: From the earliest stages of the NEPA process, all agencies with jurisdiction over mining operations or affected resources**

should be required to cooperate effectively in the scoping, preparation, and review of environmental impact assessments for new mines. Tribes and nongovernmental organizations should be encouraged to participate and should participate from the earliest stages.

Conclusion: Inefficiencies and time delays in the completion of environmental review under NEPA, issuance of permits, and conduct of other administrative actions unnecessarily consume the resources and time of many stakeholders. **Recommendation: BLM and the Forest Service should plan for and implement a more timely permitting process, while still protecting the environment.**

Conclusion: Misunderstandings of the term "unnecessary or undue degradation" (FLPMA, 1976 [43 U.S.C. §§7401 et seq.]) leave some BLM field staff uncertain whether the agency has the authority to protect valuable resources, such as riparian habitats, that may not be specifically protected by other laws. **Recommendation: BLM should prepare guidance manuals and conduct staff training to communicate the agency's authority to protect valuable resources that may not be protected by other laws.**

Conclusion: Forest Service permitting procedures for mineral exploration projects with limited environmental impact commonly take significantly longer than necessary. **Recommendation: Forest Service regulations should allow exploration disturbing less than 5 acres to be approved or denied expeditiously, similar to notice-level exploration activities on BLM lands.**

Conclusion: Deficiencies in both staff size and training were observed by the Committee in some offices of land management agencies. Increases in staffing and improved training should result in improved environmental protection and program efficiency. **Recommendation: BLM and the Forest Service should carefully review the adequacy of staff and other resources devoted to regulating mining operations on federal lands and, to the extent required, expand and/or reallocate existing staff, provide training to improve staff capabilities, secure supplemental technical support from inside and outside the agencies, and provide other support as necessary.**

REGULATORY GAPS

Although improvements in implementation present the greatest opportunities for improving environmental protection and the efficiency of the regulatory process, the committee also has identified the following specific issues or "gaps" in the existing regulations:

Conclusion: Financial risks to the public and environmental risks to the land exist whenever secure financial assurances are lacking. **Recommendation: Financial assurance should be required for reclamation of disturbances to the environment caused by all mining activities beyond those classified as casual use, even if the area disturbed is less than 5 acres.**

Conclusion: Some small mining and milling operations present environmental risks and potential financial liabilities for the public. These exposures are small by comparison to large operations, but as currently regulated they constitute a disproportionate share of the problems for the land management agencies. **Recommendation: Plans of operations should be required for mining and milling operations, other than those classified as casual use or exploration activities, even if the area disturbed is less than 5 acres.**

Conclusion: Current regulations do not provide land management agencies with straightforward procedures for modification of plans of operations even with compelling environmental justification. **Recommendation: BLM and the Forest Service should revise their regulations to provide more effective criteria for modifications to plans of operations, where necessary, to protect the federal lands.**

Conclusion: Federal criteria do not distinguish between temporarily idle mines and abandoned operations. This distinction is important because mines that become temporarily idle in response to cyclical metal prices and other factors need to be stabilized but not reclaimed, whereas mines that are permanently idle need to be reclaimed. **Recommendation: BLM and the Forest Service should adopt consistent regulations that a) define the conditions under which mines will be considered to be temporarily closed; b) require that interim management plans be submitted for such periods; and c) define the conditions under which temporary closure becomes permanent and all reclamation and closure requirements must be completed.**

Conclusion: Current regulations discourage reclamation of abandoned mine sites by new mine operators. New mineral deposits are commonly found at the sites of earlier mines. Even though the operator of a new mine might volunteer to clean up previous degradation, the long-term liability acquired under current regulations can be significant. As a result, non-taxpayer supported reclamation opportunities are missed and undisturbed lands may be preferentially disturbed for new mining sites. **Recommendation: Existing environmental laws and regulations should be modified to allow and promote the cleanup of abandoned mine sites in or adjacent to new mine areas without causing mine operators to incur additional environmental liabilities.**

Conclusion: Post-mining land use and environmental protection are inadequately addressed by both agencies and applicants. The regulations and plans of operation generally specify what actions will be taken to protect water quality and what surface reclamation is to be performed for closure. However, there is inadequate consideration of protection of the reclaimed land from future adverse uses; of very long-term or perpetual site maintenance; or of rare, but inevitable, natural emergencies. **Recommendation: BLM and the Forest Service should plan for and assure the long-term post-closure management of mine sites on federal lands.**

Conclusion: Federal land management agency representatives are inconsistent in their understanding of their enforcement authority and tools. This results from uncertain interpretations of the statutes and regulations, inadequate staff training, and deficiencies in the tools themselves. **Recommendation: Federal land managers in BLM and the Forest Service should have both (1) authority to issue administrative penalties for violations of their regulatory requirements, subject to appropriate due process, and (2) clear procedures for referring activities to other federal and state agencies for enforcement.**

RESEARCH NEEDS

Successful environmental protection is based on sound science. Improvements are needed in the development of more accurate predictive models and tools and of more reliable prevention, protection, reclamation, and monitoring strategies at mine sites. The science base is far from complete, and environmental protection requires that improvements continue to be devised. Some of the most important environmental concerns at hardrock mining sites are those related to long-term water quality and water quantity, which affect riparian, aquatic biological, groundwater, and surface water resources. A broadly coordinated, national research effort is needed to guide future development and to create improved methods for predicting, measuring, and mitigating environmental impacts related to hardrock mining. **Recommendation: Congress should fund an aggressive and coordinated research program related to the environmental impacts of hardrock mining.**

CONCLUDING PERSPECTIVES

Conditions are changing for regulations and mining. Technology, social values, the economy, and scientific understanding change continually. Thus, environmental regulations applicable to mining will be most effective if they use these changes to improve environmental protection. Similarly, the mining industry should benefit through lower operating cost and greater environmental protection. Therefore, a regulatory system that is adaptive to change will serve the public, the environment, and industry best.

Portions of the public and the mining industry have little confidence in the propriety or fairness of the regulatory and permitting system. Some members of the public perceive that regulators work too closely with the companies and permit operations without sufficient environmental safeguards. Conversely, some mining operators experience delays that they perceive to be caused, in part, by members of the public who seek to forestall mining through the permitting and regulatory processes. Lack of confidence in the regulatory and permitting system can lead to delays and higher costs for industry, regulatory agencies, and the public, and can also limit opportunities for improving environmental protection.

The Bureau of Land Management and the Forest Service need not have identical regulations, but some changes are warranted. The two agencies have broadly similar land management mandates. There are, however, some differences in the kinds of lands they manage, in their specific responsibilities, and in their organization. Whereas some of the Committee's recommendations would make the agencies' approaches to regulating hardrock mining more similar, the Committee is not suggesting that uniformity in all aspects is necessary.

All of the Committee's conclusions and recommendations, taken together, summarize the Committee's views of the actions needed to coordinate federal and state mine reclamation, operations, and permitting requirements and programs. Some of the recommendations will require congressional action and some will require changes in federal regulations. Still others will require changes in the implementation of existing regulations and programs. Adopting these recommendations will improve environmental protection and reclamation of hardrock mining on federal lands, as well as the efficiency of the permitting process.

1

Introduction

In late 1998 Congress asked the National Research Council (NRC) to conduct a study of the regulatory framework that applies to hardrock mining on federal lands. The study is particularly timely in that the Bureau of Land Management (BLM) has proposed to revise its regulations for mineral exploration and development conducted under the authority of the General Mining Law of 1872 (30 U.S.C. §22 et seq.). The BLM regulations apply to hardrock minerals—such as those bearing gold, silver, copper, and uranium—on 260 million acres of federal lands managed by BLM throughout the western United States. The Forest Service manages the national forests on which mineral exploration and mining occur and implements its own regulations for managing these activities.

As specified by Congress,[1] "the study shall identify and consider:

(A) the operating, reclamation and permitting requirements for locatable minerals mining and exploration operations on federal lands by federal and state air, water, solid waste, reclamation and other environmental statutes, including surface management regulations promulgated by federal land management agencies and state primacy programs under applicable federal statutes and state laws and the time requirements applicable to project environmental review and permitting;

(B) the adequacy of federal and state environmental, reclamation and permitting statutes and regulations applicable in any state or states where mining or exploration of locatable minerals on federal lands is occurring, to prevent unnecessary or undue degradation; and

(C) recommendations and conclusions regarding how federal and state environmental, reclamation and permitting requirements and programs can be coordinated to ensure environmental protection,

---

[1] Department of the Interior and Related Agencies Appropriations Act, 1999 P.L. 105-277, Division A, Title I, Sec. 120, enacted October 21, 1998

increase efficiency, avoid duplication and delay, and identify the most cost-effective manner for implementation."

To conduct the study the NRC established the Committee on Hardrock Mining on Federal Lands in January 1999. The Committee conducted an evidence-based analysis, taking account of scientific and technical knowledge. It was not asked to address the issue of comprehensive reform of the General Mining Law of 1872 or general problems related to abandoned mine lands. Because the Committee's charge indicated that the baseline for the study was the existing regulatory framework rather than proposed revisions to that framework, the Committee did not focus on the BLM proposal to revise its regulations governing hardrock mining; the Committee did, however, receive briefings on that proposed rule making.

The Committee concentrated on the intersection of three elements that form the context of hardrock mining on federal lands:

(1) minerals development and the factors that drive it; (Introduction and Appendix A)
(2) the natural environment and how it can be affected by hardrock mining (Introduction and Appendix B); and
(3) federal and state laws and regulations (Chapter 2 and Appendix C).

This federal lands context is governed by the General Mining Law of 1872 (30 U.S.C. §22 et seq.). The law defines the system of open access to hardrock minerals on federal lands of the western United States that are not withdrawn from mineral entry. The law allows any person to stake a claim on these lands and thereby to obtain the exclusive right to extract the minerals thereon without payment of royalty to the United States and without acquiring title to the land itself. The regulation of hardrock mining and exploration operations on these "unpatented" mining claims, where title to the land remains with the United States, is the focus of this study.

A mining claimant may also obtain title to (patent) the lands by proving the location of a valuable mineral deposit on the mining claim, among other requirements, and paying the United States the statutory price ($5.00 per acre for lode claims and $2.50 per acre for placer claims). The result of over a century of mineral patenting and other conveyances of federal lands is, in many parts of the West, a patchwork of intermingled privately owned lands; state-owned lands; and federally owned lands, including unpatented mining claims. A large mine may occupy a mixture of these land ownerships, each subject to different regulatory and land management requirements.

For the purposes of this study, we use the term "hardrock minerals" as a synonym for "locatable minerals," which is a legal term not widely used in the technical literature. Sidebar 1-1 provides a definition of locatable minerals and distinguishes locatable minerals from leasable and saleable minerals, which are also legal terms that are not widely used in the technical literature. Examples of the three groups of minerals are provided in Sidebar 1-1. Some minerals are either locatable or saleable, depending on whether they are common or uncommon varieties as defined by the Common Varieties Act of 1955 (30 U.S.C. §601 et seq.) (Sidebar 1-2). For the purposes of this report, the phrase "federal land management agencies" refers to BLM and the Forest Service.

POLICY CONTEXT

The BLM first implemented its hardrock mining surface management regulations in 1981 (codified at 43 CFR 3809; the "3809 regulations"). The preamble to the 1981 regulations stipulates that the regulations would be reviewed for possible revisions in three years. The review was postponed and was not addressed until 1989, when the BLM established a task force to address significant hardrock mining issues. In 1991 the BLM formed another review task force, but in 1993 it decided to put the review on hold because of anticipated congressional amendments to the Mining Law of 1872.

The House of Representatives and the Senate passed differing bills to reform the Mining Law of 1872 during the 103rd Congress in 1993-94. The congressional conference committee convened to reconcile the House and Senate bills, but failed to reach agreement before Congress adjourned in the fall of 1994.[2]

In January 1997 the Secretary of the Interior directed the BLM to restart the process for revising its hardrock mining regulations. Because of the time that had passed since the 1991 effort, the BLM started a new public participation process in early 1997. An agency task force with expertise in the program was created to coordinate public involvement, develop regulation options, and oversee changes in the regulations. In early 1997 BLM revised the 3809 regulations for reclamation bonding. The new bonding regulation was challenged in court. In May 1998 the District Court ruled against BLM and the 1997 bonding regulation is no longer in effect.[3]

---

[2] The Mining Law of 1872 has been the topic of intermittent debates in Congress since its passage. One history of the legal aspects of mineral development says, "The history of public mining law in this country has yet to record a plateau of comparative quietude" (Swenson,1968; see also Leshy, 1987).

[3] *Northwest Mining Association v. Babbitt,* No. 97-1013, D.D.C., May 13, 1998.

SIDEBAR 1-1 Locatable, Leasable, and Saleable Minerals

Locatable Minerals

A legal term that, for federal lands in the United States, defines a mineral or mineral commodity that is acquired through the General Mining Law of 1872, as amended (30 USC 22-54, 161, 162, 661-615). These are the base and precious metal ores, ferrous metal ores, and certain classes of industrial minerals. Acquisition is by staking a mining claim (location) over the deposit and then acquiring the necessary permits to explore or mine. Examples of locatable minerals include, but are not limited to, gold, silver, platinum, copper, lead, zinc, magnesium, nickel, tungsten, bentonite, barite, feldspar, fluorspar, uranium, and uncommon varieties of sand, gravel, and dimension stone.

Leasable Minerals

A legal term that, for federal lands or a federally retained mineral interest in lands in the United States, defines a mineral or mineral commodity acquired through the Mineral Land Leasing Act of 1920, as amended, the Geothermal Steam Act of 1970, as amended, or the Acquired Lands Act of 1947, as amended. These Acts are found in Title 30 of the United States Code—Mineral Lands and Mining. Acquisition is by application for a government lease and permits to mine or explore after lease issuance. Examples of leasable minerals include oil, gas, coal, oil shale, sodium, potash, phosphate, and all minerals within acquired lands.

Saleable Minerals

A legal term that, for federal lands, defines mineral commodities sold by sales contract from the federal government. The applicable statute is the Mineral Materials Sale Act of 1947, as amended (30 USC 602-604, 611-615). Saleable minerals are generally common varieties of construction materials and aggregates, such as sand, gravel, cinders, roadbed, and ballast material.

Source: *Dictionary of Mining, Mineral, and Related Terms*, 2nd edition, American Geological Institute, 1997.

SIDEBAR 1-2 Common and Uncommon Variety Minerals

Common Variety Minerals
Stone, gravel, pumice, pumicite, and cinders that, though possibly having value for trade, manufacture, the sciences, or the mechanical or ornamental arts, do not have a distinct, special value for such use beyond normal uses. On federal lands such minerals are considered saleable and are disposed of by sales or by special permits to local governments.

Uncommon Variety Minerals
On the federal lands stone, gravel, pumice, pumicite, and cinder deposits that have distinct and special properties making them commercially valuable for use in a manufacturing, industrial, or processing operation. Such minerals are locatable under the Mining Law of 1872, as amended. In determining a deposit's commercial value, the following factors may be considered: quality and quantity of the deposit, geographic location, accessibility to transportation, and proximity to market or point of use.

Source: Draft Environmental Impact Statement, Surface Management Regulations for Locatable Minerals Operations, U.S. Department of the Interior, BLM, February, 1999, p. G-5, G-26.

In late October 1998 Congress asked the NRC to conduct an independent study on mining of hardrock minerals on federal lands, including consideration of the adequacy of existing environmental and reclamation requirements to prevent "unnecessary or undue degradation" and the time requirements applicable to project environmental review and permitting. Congress instructed that the study was to be completed by July 31, 1999. The legislation also prohibited the Secretary of the Interior from promulgating final regulations to change the BLM regulations found at 43 CFR 3809 prior to September 30, 1999.

In February 1999 the BLM published its proposed revisions to the 3809 regulations and a draft environmental impact statement (EIS). The BLM held public meetings on the proposed regulations and draft EIS in 13 cities in March and April 1999. The deadline for providing comments to BLM on these documents was May 10, 1999.

In May 1999 Congress passed legislation requiring the Secretary of the Interior to provide a period of not less than 120 days for accepting public comment on the BLM's proposed rule after the Committee on Hardrock

Mining on Federal Lands submits its report to the appropriate federal agencies, the Congress, and the governors of the affected states.

Although the Committee was aware of the proposed changes to BLM's regulations, it has focused its review on the existing regulatory framework. The Committee was informed that the Forest Service also was considering revisions to its regulations (36 CFR 228) for hardrock mining on western national forest lands; however, the Committee considered only the existing Forest Service regulatory framework.

APPROACH TO THE STUDY

The Committee on Hardrock Mining on Federal Lands consists of 13 members with a broad range of expertise and experience (Appendix H). The Committee operates under the Committee on Earth Resources, which is under the aegis of the Board on Earth Sciences and Resources within the NRC's Commission on Geosciences, Environment, and Resources.

The study was conducted on a fast-track schedule, as requested by Congress. The Committee conducted 5 meetings, three field trips, and three public participation forums in 5 months (Appendixes F and G). The committee meetings were held in Washington, D.C. (two meetings), Denver, Reno, and Spokane. The committee meetings featured 60 speakers from federal and state agencies, environmental organizations, industry, and other institutions. The public participation forums included presentations from 37 individuals (Appendixes F and G).[4]

The Committee conducted field trips to large mines, small mines, and exploration sites on lands administered or formerly administered by BLM and the Forest Service: Cripple Creek and Victor Mine (Colorado), Sleeper Mine (Nevada), Gold Quarry Mine (Nevada), Flatrock Claims (Washington), and exploration sites on Forest Service lands in the Sullivan Ranger District (Washington). Subsets of the Committee visited other mines and exploration sites in Alaska, California, and Nevada.

To supplement its analysis of the existing panoply of federal and state statutes and regulations, the Committee sought assistance from 12 western states. The Committee sent letters to state regulatory agencies requesting information about violations of state environmental regulations applicable to hardrock mining activities on federal lands during the last 5 years. Information supplied by the state agencies ranged from minimal to comprehensive. A

---

[4] Copies of approximately 250 documents reviewed by the Committee can be requested from the Public Access Records Office of the NRC's library (through the National Academies World Wide Web site at www4.national-academies.org/cp.nsf).

limited number of examples of recent environmental impacts at hardrock mine sites are presented in Appendix B. One of the Committee's recommendations addresses the importance of creating a comprehensive management information system that tracks compliance and communicates this information to all stakeholders.

The study of hardrock mining on federal lands builds upon previous NRC reports on mining issues. It has been almost 20 years since the NRC's Committee on Surface Mining and Reclamation (COSMAR) published its report on *Surface Mining of Non-Coal Minerals* (NRC, 1979) pursuant to a congressional directive in the Surface Mining Control and Reclamation Act of 1977. In the two decades since the issuance of that report there have been significant advances in mining technology. Moreover, in the intervening years, many states have enacted and implemented extensive regulatory programs to address the environmental impact of hardrock mining.

More recently, the NRC issued a report entitled *Competitiveness of the U.S. Minerals and Metals Industry* (NRC, 1990). That report addresses the U.S. minerals and metals industry in a changing global context and discusses trends in U.S. production and consumption of metals. It finds that the United States is among the world's largest consumers of nearly every metal, much of which is imported. Moreover, it observes that, although the U.S. share of the world market for most major metals has slipped, the United States is among the world's largest producers of many important metals and still has substantial reserves. According to recent mine production data compiled by the U.S. Geological Survey (1999), the United States is the world's largest producer of molybdenum and second largest producer of gold, silver, and copper.

FEDERAL LANDS AND HARDROCK MINING

Federal lands have been an important source of minerals since the early part of the nineteenth century. Although most of what were once federal lands between the Appalachian and Rocky mountains are now under non-federal ownership, the remaining federal lands in the western states, including Alaska, continue to provide a large share of the metals and hardrock minerals produced in this country. The federal lands generally available for exploration for hardrock minerals are part of the original federal public domain and are now under the management of the BLM, an agency of the Department of the Interior, and the Forest Service, an agency of the Department of Agriculture.

Lands administered by the BLM and Forest Service comprise 38% of the combined land area of the 12 western states (Table 1-1). The combined

Table 1-1 Public Lands Administered by BLM and the Forest Service

State	BLM Lands (1) Acres	% of state acreage	Forest Service Lands (2) (3) Acres	% of state acreage	Sum of BLM and Forest Service Lands Acres	% of state acreage	Federal Total Acres	% of state acreage	Total State Acreage (4)
Alaska	86,908,060	23.8	21,969,321	6.0	108,877,381	29.8	171,787,844	47.0	365,481,600
Arizona	14,220,457	19.6	11,251,701	15.5	25,472,158	35.0	31,336,526	43.1	72,699,000
California	14,556,074	14.5	20,647,142	20.6	35,203,216	35.1	44,757,474	44.7	100,206,720
Colorado	8,260,306	12.4	14,508,108	21.8	22,768,414	34.2	24,128,982	36.3	66,485,760
Idaho	11,775,052	22.2	20,460,774	38.7	32,235,826	60.9	32,991,905	62.3	52,933,120
Montana	6,259,211	6.7	16,877,005	18.1	23,136,216	24.8	25,485,347	27.3	93,271,040
Nevada	47,841,264	68.1	5,823,676	8.3	53,664,940	76.4	56,081,559	79.8	70,264,320
New Mexico	12,541,069	16.1	9,326,935	12.0	21,868,004	28.1	26,217,230	33.7	77,766,400
Oregon	16,145,615	26.2	15,656,351	25.4	31,801,966	51.6	31,809,283	51.6	61,598,720
Utah	22,832,630	43.3	8,112,730	15.4	30,945,360	58.7	33,898,255	64.3	52,696,960
Washington	370,110	0.9	9,177,071	21.5	9,547,181	22.4	11,938,958	28.0	42,693,760
Wyoming	18,373,492	29.5	9,247,742	14.8	27,621,234	44.3	30,878,166	49.5	62,343,040
Sub Total	260,364,677	22.3	165,082,211	14.1	425,446,888	36.4	563,080,110	48.2	1,167,797,880
U.S. Total	261,614,888	11.5	191,785,560	8.4	453,400,448	20.0			2,271,343,360
12 Western States as a % of U.S. total		99.5		86.1		93.8		51.4	

NOTES: (1) BLM, 1999a, p. 78. These data exclude Land Utilization Project lands, to which the 3809 regulations do not apply. Land Utilization Project lands comprise 2.3 million in 9 western states, including 1.8 million acres in Montana. BLM's 3809 regulations also apply to Stockraising Homestead Act Acreage, which include 67.1 million acres in the 12 western states, including 18.1 million acres in Wyoming and 15.6 million acres in New Mexico (2) Forest Service, 1998, pp.14-35. (3) GAO, 1996, p.11. (4) BLM, 1999a, p. 78.

acreage of BLM and Forest Service lands range from about 76% of Nevada to about 23% of Washington (BLM, 1999b). In addition to the BLM and Forest Service lands, sometimes called the "multiple use" lands because they serve many uses, the federal government also has large land holdings in the West devoted to specific uses. These include the national parks and national wildlife refuges, which are managed by the National Park Service and Fish and Wildlife Service in the Department of the Interior, military lands managed by Department of Defense agencies, and lands managed by the Department of Energy.

The 12 western states contain 99% of the lands administered by BLM and 85% of the lands administered by the Forest Service. BLM lands in these 12 states total about 260 million acres and Forest Service lands about 163 million acres (Table 1-1). These vast areas compare with an estimated 15,000 acres of currently active notice-level mining operations (less than 5 acres each) and 134,000 acres in currently active plan-level mining operations on BLM lands (BLM, 1999a, p. 86-87). Based on these data, approximately 0.06% of BLM lands are affected by currently active plan-level and notice-level mining activities (see Sidebar 1-3 for a discussion of categories of mining activities on BLM lands). However, impacts to water quality, vegetation, and aquatic biota from both active and historic mining often extend beyond the immediate area of the mine site. In addition, there are over 200,000 inactive and abandoned mines, many of which are on federal lands (EPA, 1997b). Not all BLM and Forest Service lands are available for mineral prospecting and mining. Various categories of these lands have been withdrawn from mineral entry to meet a variety of public purposes. Chief among these are congressionally designated wilderness areas. Other lands have been withdrawn administratively to protect relatively large areas being considered for possible designation as wilderness, areas used for military facilities, and smaller areas for recreation and for other facilities. The Committee requested information about the acreage of BLM and Forest Service lands that are withdrawn from mineral entry, but these data were not readily available.

It is difficult to obtain accurate information about the level of mining activities on federal lands, because neither BLM nor the Forest Service systematically collect data on the value of hardrock minerals produced on their lands. The BLM reports the number of plans and notices of mining operations filed with the agency on an annual basis. Time series data are provided in Figures 1-1 and 1-2. These data indicate that the number of plans of operations filed with the BLM has decreased roughly by 50% since 1992 (Figure 1-1). The number of notices has fallen by a greater percentage, apparently due in large part to the imposition of a $100 holding fee per mining claim per year instead of the previous requirement that claim holders conduct $100 of assessment work per year (Figure 1-2). Declining levels of exploration and lower metals prices may also account in part for recent declines in filings of plans of operations.

SIDEBAR 1-3 Categories of Mining Activities on BLM Lands

Casual Use

Mining activities that only negligibly disturb federal lands and resources. Casual use does not include the use of mechanized earth moving equipment or explosives or the use of motorized equipment in areas closed to off-road vehicles. Under casual use, operators do not have to notify BLM, and operations do not need to be approved. But operations are subject to monitoring by BLM to ensure that federal lands do not undergo unnecessary or undue degradation. Casual use operations must be reclaimed.

Notice-level Operation

A mining or exploration operation involving more than casual use but requiring that the operator submit only a notice rather than a plan of operations. A notice is the notification a mining operator must submit to BLM of the intention to begin an operation that will disturb 5 acres or less in a year in a mining claim or project area. The intent of a notice is to permit operations with limited geographic disturbance to begin after a quick review for potential resource conflicts and to eliminate the need for federal action. A notice requires no special forms, but an operator must submit specific information. BLM must complete its review of the notice within 15 calendar days of its receipt unless more information is required to determine whether the operation would cause unnecessary or undue degradation.

Plan of Operations

A plan for mining exploration and development that an operation must submit to BLM for approval when more than 5 acres a year will be disturbed or when an operator plans to work in an area of critical environmental concern or a wilderness area. A plan of operations must document in detail all actions that the operator plans to take from exploration through reclamation.

Source: Draft Environmental Impact Statement, Surface Management Regulations for Locatable Minerals Operations, U.S. Department of the Interior, BLM, February, 1999, p. G-4, G-16, G-18.

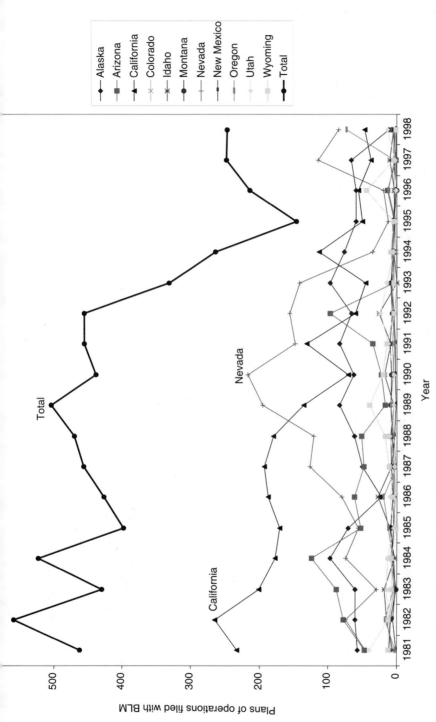

FIGURE 1-1 Plans of operations filed with BLM for mining or exploration activities on BLM lands. Source: BLM, 1992, 1994, 1999b.

22

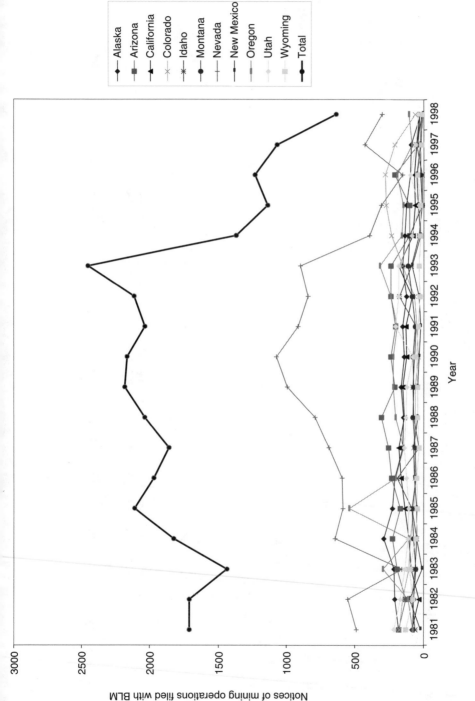

FIGURE 1-2 Notices of intent filed with BLM for exploration or mining activities on BLM lands. Source: BLM, 1992, 1994, 1999b.

HARDROCK MINING

Hardrock mining occurs where minerals are concentrated in economically viable deposits. Ore deposits form as variants of such geologic processes as volcanism, weathering, and sedimentation operating with an extraordinary intensity. Ore deposits typically are parts of large-scale (several miles across and perhaps just as deep) ore-forming systems in which many elements, not just those of economic interest, have been enriched. Known ores constitute less than 0.01% of the metal content of the upper 1 km of continental crust. Thus, mines can only be located in those few places where economically viable deposits have been formed and discovered (see Sidebar 1-4).

Many hardrock commodities are associated with magmatic and hydrothermal processes (Guilbert and Park, 1986). These processes, in turn, are associated with modern or ancient mountain belts. The abundant igneous rocks and associated hydrothermal systems and the mountainous or sparsely vegetated terrain make the West the location of most hardrock mines in the United States. Some of these same areas also are valued for aesthetic and cultural reasons, which creates potential for conflict among uses of the land. While society requires a healthy environment, it also requires sources of materials, many of which can be supplied only by mining.

The mining process consists of exploration, mine development, mining (extraction), mineral processing (beneficiation), and reclamation (for closure), which are discussed briefly in the following section. A more detailed discussion of the mining process can be found in Appendix A.

Exploration

The primary objective of exploration is to find an economic mineral deposit. The initial step in exploration is prospecting, which is the search for indications of a mineral deposit of potential significance. The objective is to define a target that suggests the occurrence of a mineral deposit worthy of subsequent testing. Target identification may result from relatively unobtrusive approaches, such as detailed surface geologic mapping, sampling of outcrops, and observation of alteration patterns, or from more technical methods, including various types of geochemical sampling, geophysical techniques, satellite remote sensing, and other sophisticated methods for identifying the often subtle expressions of deeply buried mineral deposits.

The second phase of exploration involves target testing. The techniques vary, but they usually focus on confirming the presence of a deposit and delineating its size, shape, and composition. In most cases, the exploration tool is drilling. Drilling into the subsurface is done to determine both the lateral

SIDEBAR 1-4 How hard is it to find a mineral deposit?

The art and science of finding new mineral deposits is much better than pure luck, but it is still far from perfect. Moreover, the search for new mineral deposits is costly, time consuming, and without guarantee of success. For example, Roscoe (1971) showed that the number of mineral indications in Canada that had to be investigated to discover a significant mineral deposit was about 100 in 1951 and rose to about 1000 in 1969. There is no reason to expect that this trend has changed. Similarly, in a probabilistic analysis of exploration experience in the United States by Homestake Mining Company, Anderson (1982) concluded that from an initial sample of 1,000 reconnaissance examinations (more or less equivalent to casual use activities), 100 drillable exploration targets (roughly equivalent to notice-level activities) would emerge in which there would be a 75% chance of finding one deposit with 3 million ounces of gold. The statistics may not be quite as grim as they first appear, because there are many cases of someone with a better concept, more persistence, or luck finding an economic deposit in a prospect or worked-out mine that several companies have deemed worthless. Successful projects can be spectacularly profitable, but overall, mining has one of the lowest returns on investment of major industries (Dobra, 1997).

and vertical extent of the deposit and other important deposit characteristics, such as continuity of mineralization, grade and associated trends, mineralogical relationships and characteristics, rock types, and hydrologic information.

In general, the environmental impact of prospecting is minimal. Surface disturbance resulting from other exploration activities increases as these activities progress from preliminary to detailed. Most surface disturbance results from the construction of access roads and drill sites. Surface disturbance also may result from trenching activities across zones of mineralization or from the collection of bulk samples for metallurgical testing. The subsurface environment also may be affected by drilling, although the effects are usually minor due to the use of practices such as sealing drill holes and the proper disposal of drill cuttings. A more significant subsurface impact is often that associated with the collection of bulk samples of ore for metallurgical testing.

When exploration on federal lands indicates a valuable deposit, the prospector acquires rights, if not previously completed, to develop it by staking claims and recording them with appropriate agencies. The area required for a large mine and its facilities (e.g., waste dumps, tailings ponds) is frequently a few thousand acres. Because of intermixed ownerships, this often involves a combination of federal and private lands for a single mine.

Mine Development

When a deposit identified and defined from the exploration program has been judged economic and the required permits have been obtained, the next activity is mine development to prepare the deposit for extraction or mining. At this stage in the mining process the nature of the environmental impacts is different and their magnitude greater than those associated with exploration activities. The footprint of the mining operation is essentially defined at this point. Infrastructure (e.g., power, roads, and water) is put in place throughout the site, and planning and construction of support facilities (e.g., offices, maintenance shops, fuel bays, and materials handling systems) and the mineral processing facility is completed. Surface locations are delineated and prepared for placing waste or overburden material, heap leach piles (the chemical leaching of low-grade ores), stockpiles, and tailings impoundments.

Near-surface deposits in open-pit mines are prepared initially for production by removing the overlying waste material or overburden, which is then moved to waste dumps on the surface. Deeper deposits to be mined underground are developed initially by gaining access to the mineralization through vertical or inclined shafts or horizontal adits. Underground drifts, crosscuts, raises, and ramps are excavated to provide the access needed to mine the ore.

Mining (Extraction)

As mine development nears completion, there is a transition to the mining component, which refers to extraction of the mineralized material (and associated waste rock) as the mine becomes deeper and grows laterally. Although every ore deposit is different, most mines (both surface and underground) use the same basic operations: drilling, blasting, mucking (loading), and transporting (hauling).

Placer mining, in which gold or other minerals are extracted from stream or beach sediments by gravity separation using water, is different from typical hardrock mining and mineral processing in several respects. Most current placer mining in the United States involves mining with mechanized earth-moving equipment or suction dredging in streams. Placer mining with mechanized earth-moving equipment typically involves relocating a short (on the order of 2,000 feet) stretch of a stream, removal of the vegetative mat or soil, mining of gravels, removal of gold with sluices that separate dense from light minerals, and reclamation by replacing gravel and the vegetative mat or soil. Suction dredging pumps sediment from the stream bottom and processes it in a floating sluice, and can be as small as a "recreational" one-person operation.

Mineral Processing (Beneficiation)

Mineral processing, or beneficiation, consists of upgrading or concentrating the ore before the concentrate is transported to a smelter or refinery. It begins with crushing and grinding the ore into fine particles, thereby disaggregating ore grains and waste mineral grains. These particles are then subjected to physical and chemical processes using various chemicals and reagents to separate and concentrate the valuable minerals from the unwanted waste and other substances in the ore. Especially with copper ores, metals can be extracted on site directly through treatment of the ores by a process called solvent extraction/electrowinning. Otherwise, concentrates are reduced to metals by smelting.

Heap leaching is an increasingly common practice for the extraction of concentrated metals from ore, using cyanide solutions for gold ore and ferric sulfate/sulfuric acid solutions for low-grade copper ore. In heap leaching, ore can be leached without crushing, or it can be crushed only fine enough to allow the lixiviant (leaching solution) access to most of the mineral grains. The solution percolates through the piled ore, and the resulting metal-containing solutions are pumped to a processing facility for extraction. The fluids are then treated and recycled.

The waste or unwanted minerals (tailings) separated from the valuable minerals are routinely disposed of in a tailings pond near the mine site. The associated water is typically recycled, treated, and used in subsequent mining or processing. Tailings generally contain small amounts of the valuable mineral(s) not completely recovered during beneficiation, some unwanted or undesirable deleterious minerals, waste rock minerals, and some fluids with chemicals and metals from the separation process. Tailings from underground mines may be used to backfill underground voids. Tailings dams and ponds and leach pads are designed to high standards, but some do fail, and the release or discharge of tailings, leached rock, or pregnant leach solutions can have negative environmental impacts.

Reclamation

Reclamation minimizes the potential for future environmental damage and prepares the mining and processing site for beneficial use after mining. Common reclamation practices include reducing the slopes on the edges of waste rock dumps and heaps to minimize erosion; capping these piles and tailings piles with soil; planting grasses or other plants that will benefit wildlife or grazing stock and help prevent erosion; directing water flow with French drains and other means to minimize the contact of meteoric water with potentially acid-generating sulfides in the dumps, heaps, and tailings piles; removing buildings; and eliminating roads to minimize unnecessary future entry by vehicles. A mine is not closed until reclamation is complete, but as discussed in this report, in some circumstances reclamation may never be accomplished fully, and long-term monitoring will be necessary.

POTENTIAL ENVIRONMENTAL IMPACTS OF HARDROCK MINING

From exploration through post-closure, hardrock mining has the potential to cause environmental impacts. In addition to the obvious disturbance of the land surface, mining may affect, to varying degrees, groundwater, surface water, aquatic biota, aquatic and terrestrial vegetation, wildlife, soils, air, and cultural resources. Actions based on environmental regulations may avoid, limit, control, or offset many of these potential impacts, but mining will, to some degree, always alter landscapes and environmental resources. Regulations intended to control and manage these alterations of the landscape and the environment in an acceptable way are generally in place and are updated as new technologies are developed to improve mineral extraction, to reclaim mined lands, and to limit environmental impacts. Therefore, the committee emphasizes that these potential impacts will not necessarily occur, and when they do, they will not occur with the same intensity in all cases. Many of the impacts discussed in this section of the report would violate current regulatory requirements and standards and would be subject to enforcement actions. Nevertheless, an understanding of the potential for mining to cause environmental impacts is essential to assessing and improving the regulation of hardrock mining on federal lands. A more detailed discussion with documentation of potential environmental impacts of hardrock mining can be found in Appendix B. In this section the Committee provides a brief overview of some of these potential impacts.

Potential Impacts on Water Quality

If not mitigated through regulation and prevention strategies, hardrock mining can cause a number of significant long-term impacts to surface water and groundwater quality. Hardrock mining of metalliferous deposits can release to the environment metals, metalloids, sulfate, cyanide, nitrate, suspended solids, and other chemicals. Acid drainage has been considered one of the most significant potential environmental impacts at hardrock mine sites (University of California, 1988; EPA, 1997b). Increased awareness of this potential problem has improved methods that prevent and treat acid drainage at the source, thus minimizing impacts to water quality, aquatic biota, and other resources. Pit lakes that develop when surface mines fill with water can cause alterations of pre-mining water quality and quantity. In addition, the discharge of groundwater withdrawn to dewater pits can cause a variety of environmental problems because these waters sometimes contain higher levels of total dissolved salts, metals, and other chemicals than the streams into which they are discharged.

Potential Impacts on Aquatic Ecosystems

If the water quality problems discussed above occur at a hardrock mining site, they can have significant impacts on aquatic biota. Most metals and cyanide, even at low concentrations, are toxic to aquatic life. In addition to their toxic effects, low concentrations of some metals (even below the chronic criteria) may cause fish to avoid certain waters and impair their growth. This can be an issue for anadromous (migrating from fresh to salt water) fishes such as certain threatened and endangered salmonids, which may avoid streams contaminated with metals (even at low concentrations), resulting in the elimination of that species from the watershed.

Mechanized placer mining in active streams and suction dredge mining disturb to some degree streambed sediments, which provide habitat for macroinvertebrates and spawning habitat for salmonids. A streambed disturbed in this way may nearly return to its original characteristics after springtime high flows. During low-precipitation years, however, the streambed may remain unsuitable for aquatic life habitat until high flows return it more to its original characteristics.

Potential Impacts on Water Quantity

Changes in surface water and groundwater quantity that result from hardrock mining activities can have significant environmental impacts if not mitigated. For example, under certain conditions, surplus water accumulated from mine dewatering that is discharged into nearby streams can disrupt riparian ecosystems, erode or straighten stream channels, or dislodge aquatic biota. Groundwater withdrawal for mineral processing and to prevent filling of open pits and underground mines can affect local and regional groundwater quantities and levels. The presence of pit lakes can affect regional aquifers if they act as a permanent evaporative sink.

Lowering of alluvial aquifers because of extensive groundwater withdrawals can directly affect riparian vegetation that depends on this water source, even some distance from a mine. Wetlands, like riparian ecosystems, are dependent on a continuous supply of water, and therefore any change in regional hydrology may affect wetlands, especially in the arid West. Many arid region wetlands develop at springs, and such wetlands and spring pools often support threatened or endangered species (e.g., pupfish). These spring wetlands may be very sensitive to changes in the hydrologic head of the regional or local aquifer resulting from groundwater withdrawal by mines or other uses (e.g., agriculture or municipal uses).

Potential Impacts on the Landscape and Terrestrial Ecosystems

In addition to those potential environmental impacts related to water quality and quantity, mining activities can have direct impacts on the landscape and terrestrial ecosystems. Of special concern are impacts on relatively pristine ecosystems in remote mountainous terrains where human activity has caused little disruption. Early exploratory surveying by qualified geologists has no more impact on these remote areas than geological mapping or casual recreation, but exploration activities designed to validate a mineral deposit are potentially disruptive. For example, exploration roads, and even tracks from soft-tire vehicles, may impede migration of small mammals and change behavioral patterns of larger animals. Permanent long-distance haul roads or railroads for mining purposes have the same potential impact on animal behavior as do roads and railroads used for other purposes. That is, they may alter migration patterns by creating barriers and fragmenting animal territories. Terrestrial ecosystems also can be disrupted by the introduction of non-native plant species during reclamation activities or by dispersal of seeds by vehicular traffic.

The use of valleys for placing waste rock, leach pads, or tailings impoundments has the potential to disrupt riparian ecosystems that depend on stream flows and shallow groundwater. Although there is greater awareness today of the ecological impacts of using valleys in this way, valleys are still used in some cases for the placement of mining facilities. Valley fills destroy vegetation found within the valley and the riparian vegetation cannot be restored.

Terrestrial wildlife, waterfowl, and migratory birds may be affected by hardrock mining through bioconcentration. Plants growing in contaminated sediments accumulate metals in tissues. Aquatic plants, macroinvertebrates, and fish can develop elevated levels of metals and other contaminants in their tissue. Consumption of these organisms by wildlife and birds continues the bioconcentration process, potentially creating toxic levels of metals and other chemicals in organisms that are high on the food chain.

Some environments created by mining activities have the potential to benefit wildlife. For example, abandoned mine tunnels may be used by bat communities. Reclamation of waste rock sites and other surface disturbances may create extensive areas of forage that attract some species.

Other Potential Environmental Impacts

If not mitigated, hardrock mining activities can lead to other potential environmental impacts, including pollution of air and soils with metals and sulfur dioxide from smelter emissions; generation of fugitive dust; alteration of soils, including increased erosion; and generation of noise, which can disrupt wildlife and modify human behavior.

Cumulative Effects

Although a single mining operation may create its own set and degree of environmental impacts, a regional concentration of mines may pose problems of cumulative impacts. For example, groundwater withdrawal at a single mine has the potential to create a deep cone of depression in the local aquifer. As the cone expands over time, it may join those created by neighboring mines and lower the regional water table, which in turn may decrease or terminate flow in streams and springs some distance from the mines. Similarly, contaminated drainage or leachate from waste rock dumps, heap leach pads, or tailings ponds at a number of mines may cause environmental problems even when a single mine would not be sufficient to lower stream water quality below acceptable concentration levels. Other issues involving cumulative impacts include the potential for contamination of groundwater aquifers, fugitive dust and air pollution from tailings and road surfaces, smelter emissions, and landscape degradation from large mine operations in the same general area. Although most of the possible environmental effects of mining discussed in the previous sections are in the context of individual mine operations, the potential for cumulative impacts also should be considered.

Long-Term Monitoring

Monitoring of environmental conditions and responses to human activities is needed to measure changes in the environment, and to determine the effectiveness of mitigation procedures. Monitoring is applicable to the various stages of mining from exploration, through development and extraction, to closure, reclamation, and post-closure. Baseline monitoring is essential to establish conditions prior to any mining activity and to provide the conditions against which future monitoring data can be compared. Monitoring during mining should address the implementation and effectiveness of environmental controls and compliance with regulations. Long-term monitoring can be used to address the effectiveness of mine closure activities, to validate predictive models, and to follow long-term responses to post-closure conditions. Long-term monitoring also may capture unexpected events that could alter environmental conditions. Long-term monitoring must be properly designed to measure those attributes that offer useful data on changes and/or sustainability of resources affected by mining activity and post-closure conditions.

CHANGING CONTEXTS

Federal and state agencies have been active in developing the existing regulatory framework, which is focused on managing the types of mining activities common during the last two decades. However, mining is not a static activity. Substantial changes are occurring in mining and environmental technologies. Market conditions for many minerals vary significantly over time, and changes in the regulatory environment are themselves affecting the nature of new mining proposals. Socio-economic changes in U.S. mining regions and the nation as a whole may have a profound impact on future mining. Regulatory and land management agencies will need to attend to these issues if the agencies are to be sufficiently flexible, technically proficient, and able to adapt and respond to changes.

Changes in Technology

All phases of mining have undergone substantial technological change over the last few decades. Exploration uses remote-sensing techniques, improved conceptual models, new geochemical and geophysical instrumentation, statistical analyses and visualization of large data sets, and global positioning system capabilities, many of which were unknown two decades ago. Larger excavation equipment has allowed ore to be extracted from larger and deeper pits and has made open-pit mining feasible in areas where it would not have been possible previously. Cyanide-leaching technologies have largely replaced traditional beneficiation and some types of gold ore milling now make it economical to process ores of much lower grade.

Similar changes can be expected in the future. With the depletion of large-scale, low-grade surface ore deposits, increased consideration is being given to developing underground mines. For example, in the United States the extraction of uranium has shifted from a traditional mining and beneficiation process to an *in situ* process that depends on injecting a solution into the ground to leach the uranium from the ore without excavating the ore deposits. The leachate is collected through a system of wells, and the mineral is extracted with little surface disturbance; the technique creates a new threat to groundwater supplies, however, and care must be taken to avoid contamination of groundwater. Such *in situ* processes have been used for years for sulfur and potash, and may become commercially viable for certain other types of ore deposits (including hardrock minerals) as well. Additional technological advances may make it possible to mine deposits that, for a variety of geological, geochemical, and environmental reasons, are considered infeasible today. In many cases technological changes in the mining industry, as in other industries, will respond to increased concern about environmental degradation.

Each change in technology presents new regulatory questions and challenges, one of which is to be able to predict the long-term effects of a new technology. The regulatory agencies will need to have the expertise to respond effectively to such changes and encourage the continued development of more environmentally friendly technologies.

Changing Market Conditions

Changes in market conditions can be even more dramatic and occur over a much shorter period than changes in technology. For example, the rising price of gold during the 1980s was a major stimulus to the rapid increase in gold mining on federal lands, but more recently (since 1992) gold prices in general have fallen as additional mines are developed in the United States and abroad and as central banks reduce their gold holdings.

Other minerals have seen similar variations in recent prices as a result of such disparate events as the discovery and development of new ore deposits; changes in the political and economic situation in producing countries; technological changes in the mineral-consuming industries that can substantially change the demand for a mineral; world economic conditions; and speculation.

Regulation in the United States has been trying to catch up with such market-driven trends as the rapid increase in gold mining during the 1980s. These up and down trends have occurred in the recent past for uranium, copper, and silver; similar trends probably will occur for the same or other minerals in the future. In one sense, rising prices present a relatively favorable economic environment for imposing new regulations, because the mining companies are eager to extract the mineral while prices are high and while they can afford the cost of increased environmental protection. With falling prices these permitting negotiations may become much more difficult, and some existing mines and mining companies may become unprofitable. One predictable result is temporary closure of existing mines until prices recover. If prices do not rebound, or if other unfavorable events transpire, temporary closures may become permanent. This means that regulatory approaches must address the challenge of falling prices to assure that resources are available to pay for site reclamation and post-closure costs.

Changes in the Regulatory Environment

The substantial changes that have taken place in the U.S. regulatory environment over the last two decades have also had a significant impact on the characteristics of the mining industry. While formerly high mineral prices may have attracted operations with insufficient capital and staying power, it is only the larger, more established firms that can afford to make the investment in time, extensive data collection, complex analyses, and expensive environmental protection measures that the state and federal regulations currently require.

This is likely to be a positive trend for the environment, for these companies have a strong ongoing interest in seeing the process work smoothly. Because they are involved for the long haul, they will go to great lengths to avoid creating problems that may cause regulators to be reluctant to approve modifications to their existing plans or proposals for new operations. They also have a strong interest in ensuring that other companies behave properly as well, because they are aware of how environmental failures can have a negative impact on their ability to operate.

On the other hand, under conditions of lower metal prices, existing operations with insufficient capital may not be able to afford the continued cost of environmental protection and may abandon their operations prior to full reclamation. If financial assurances are insufficient, the public will be responsible for reclamation and post-closure costs. Regulatory systems must be able to protect the environment under both higher and lower metal price conditions by requiring adequate financial assurances and including flexible, performance-based standards that do not stifle innovation.

The Committee received testimony from senior officers of mining companies as well as other evidence that delays and uncertainties associated with the U.S. regulatory environment are causing mining companies to replace domestic operations with overseas projects, a trend that is already strongly demonstrated in exploration. There are, of course, many other factors that affect such decisions, such as the availability of rich, easily mined deposits. This trend could be reversed by changes in the regulatory or political environment in foreign countries. Such changes, combined with the adoption of more efficient regulatory systems in the United States, could encourage companies to pursue mineral exploration opportunities in the United States more aggressively. Like mining technologies, regulatory programs evolve. A clear demonstration of this evolution is the growth of new state reclamation and mining waste statutes and regulations over the last two decades.

Socio-economic Changes

The population of many of the western mining states is increasing rapidly, often with people who come from regions where hardrock mining has not been a common activity. Consequently, new residents may have values and interests that differ substantially from those of other residents who depend on mining for jobs and other economic opportunities. In addition, the U.S. population today generally expects more from its federal lands—more recreational opportunities, more wildlife and habitat protection, more watershed protection, more timber and forage production, more historic and cultural preservation, more sensitivity to tribal concerns, as well as more mineral activities. It is likely to become increasingly difficult to find sites that will not stimulate some opposition from groups with competing values and interests in the same lands.

This array of changes in technology, mineral prices, regulatory practice, and in socio-economic factors poses a profound challenge for federal land managers.

ORGANIZATION OF THE REPORT

The Committee has organized this report to respond to the three points of the statement of task. Chapter 1 provides an introduction to the report. Chapter 2 addresses the first point regarding the current regulatory requirements for mining on federal lands and time requirements for environmental review and permitting. Chapter 3 covers the second point regarding the adequacy of current regulatory requirements to prevent unnecessary or undue degradation of federal lands. Chapter 4 speaks to the third point regarding recommendations and conclusions about how federal and state regulatory requirements and programs can be coordinated to ensure environmental protection, increase efficiency, avoid duplication and delay, and identify the most cost-effective manner for implementation.

Appendix A provides an introduction to mining, its purposes, and its potential environmental impact. Appendix B discusses in some detail the potential environmental impacts of mining. Appendix C contains compilations of federal and state permitting, operating, and reclamation requirements and supplements the presentation of regulatory requirements in Chapter 2. Appendix D addresses research needs and supplements the discussion of this topic in Chapters 3 and 4. Appendix E discusses financial assurance for mining operations. Appendixes F and G list those persons who supplied input to the Committee. Appendix H provides biographical sketches of the Committee members. Technical terms and acronyms are defined, respectively, in the glossary and list of acronyms at the end of the report.

2

Existing Regulatory Framework

The study shall identify and consider . . . the operating, reclamation, and permitting requirements for locatable minerals mining and exploration operations on federal lands by federal and state air, water, solid waste, reclamation and other environmental statutes, including surface management regulations promulgated by federal land management agencies and state primacy programs under applicable federal statutes and state laws and the time requirements applicable to project environmental review and permitting.

This chapter summarizes the federal land management standards and operating, reclamation, and permitting requirements for locatable (hardrock) minerals mining and exploration on federal lands.

STATUTORY POLICIES FOR MANAGEMENT

To assess the adequacy of regulatory requirements, the Committee examined the objectives set forth by Congress for federal land managers to apply to hardrock mining activities on federal lands.

The Federal Land Policy and Management Act of 1976 (43 U.S.C. §1701-1784) provides the Bureau of Land Management (BLM) with management standards for hardrock mining activities on public lands. Section 1701(a) states:

The Congress declares that it is the policy of the United States that—
(7) goals and objectives be established by law as guidelines for public land-use planning, and the management be on the basis of multiple use and sustained yield unless otherwise specified by law.[1]

(8) the public lands will be managed in a manner that will protect the quality of scientific, scenic, historical, ecological, environmental, air and atmospheric, water resource, and archeological values; that, where appropriate, will preserve and protect certain public lands in their natural condition; that will provide food and habitat for fish and wildlife and domestic animals; and that will provide for outdoor recreation and human occupancy and use.

(12) the public lands be managed in a manner which recognizes the Nation's need for domestic sources of mineral, food, timber, and fiber from the public lands, including implementation of the Mining and Minerals Policy Act of 1970 as it pertains to the public lands.[2]

---

[1] 43 U.S.C. §1702(c) defines multiple use as "a combination of balanced and diverse resource uses that takes into account the long-term needs of future generations for renewable and nonrenewable resources, including, but not limited to, recreation, range, timber, minerals, watershed, wildlife and fish, and natural scenic, scientific, and historical values; and harmonious and coordinated management of the various resources without permanent impairment of the productivity of the land and the quality of the environment with consideration being given to the relative values of the resources and not necessarily to the combination of uses that will give the greatest economic return or the greatest unit output."

[2] The Mining and Minerals Policy Act of 1970 states: "The Congress declares that it is the continuing policy of the Federal Government in the national interest to foster and encourage private enterprise in (1) the development of economically sound and stable domestic mining, minerals, metal and mineral reclamation industries, (2) the orderly and economic development of domestic mineral resources, reserves, and reclamation of metals and minerals to help assure satisfaction of industrial, security, and environmental needs, (3) mining, mineral and metallurgical research, including the use and recycling of scrap to promote the wise and efficient use of our natural and reclaimable mineral resources, and (4) the study and development of methods for the disposal, control, and reclamation of mineral waste products, and the reclamation of mined land, so as to lessen any adverse impact of mineral extraction and processing upon the physical environment that may result from mining or mineral activities." 30 U.S.C. 21a.

And Congress provided in section 1732(b):

> In managing the public lands the Secretary shall, by regulation or otherwise, take any action necessary to prevent unnecessary or undue degradation of the lands.

The Forest Service manages mining and its impacts under the standard of the 1897 Organic Act, which grants the Secretary of Agriculture general power to promulgate rules to regulate "occupancy and use and to preserve the forests thereon from destruction."[3] The Forest Services also operates under the National Forest Management Act (16 U.S.C. 1600-1640).

Both agencies also have management authority under the Surface Resources Act, which applies to mining claims located after 1955: "Rights under any mining claim hereafter located under the mining laws of the United States shall be subject . . . to the right of the United States to manage and dispose of the vegetative surface resources thereof and to manage other surface resources thereof."[4]

These statutory authorities find further expression in the regulations adopted by the respective agencies (43 CFR 3809 and 36 CFR 228) but the regulations do not exhaust the authority of the agencies. The statutory standards, coupled with the agencies' obligations under the National Environmental Policy Act (NEPA) to "interpret and administer" their laws in accordance with the policies set forth in NEPA,[5] make it clear that the BLM

---

[3] 16 U.S.C. §551. United States v. Weiss, 642 F.2d 296 (9th Cir. 1981) (upholding Forest Service Part 228 regulations under the Organic Act); United States v. Doremus, 888 F.2d 630 (9th Cir. 1989) (upholding Forest Service Part 228 regulations under Surface Resources Act).

[4] 30 U.S.C. §612.

[5] NEPA §102(1), 42 U.S.C. §4332(1), provides that "The Congress authorizes and directs that, to the fullest extent possible: (1) the policies, regulations, and public laws of the United States shall be interpreted and administered in accordance with the policies set forth in the Act." Section 105 of NEPA, 42 U.S.C. §4335, further provides that "the policies and goals set forth in this Act are supplementary to those set forth in existing authorization of Federal agencies." The policies are the six set out in §101(b), 42 U.S.C. §§4331(b): to "(1) fulfill the responsibilities of each generation as trustee of the environment for succeeding generations; (2) assure for all Americans safe, healthful, productive, and esthetically and culturally pleasing surroundings; (3) attain the widest range of beneficial uses of the environment without degradation, risk to health or safety, or other undesirable and unintended consequences; (4) preserve important historic, cultural, and natural aspects of our national heritage, and maintain, wherever possible, an environment which supports diversity and variety of individual choice; (5) achieve a balance between population and resource use which will permit

and Forest Service have public responsibilities that go beyond those of the state regulatory agencies seeking to protect specific environmental media. The federal agencies as land managers on the public's behalf stand in a different relationship to the land and its resources than simply as regulators of impacts. The federal land managers have a mandate for long-term productivity of the land, protection of an array of uses and potential future uses, and management of the federal estate for diverse objectives.

This relationship means that the term "regulator" does not fully describe BLM and Forest Service responsibilities when dealing with mining activities on federal lands. It also means that these agencies are not merely landholders. They are both landholders and regulators, with a set of statutory management standards. Further, they must serve a constituency almost always described in national terms—"the nation's needs," "all Americans," "future generations."

These factors explain in part why federal officials told the Committee that even in those states where working relationships and regulatory coordination were almost uniformly described as positive and productive, they were concerned about their ability to rely entirely on state regulators to address both short-term and long-term land use implications of mining operations on federal lands. Concerns surfaced in such areas as the ability to track and respond quickly to expansions in disturbed areas at small mines and to address post-mining conditions at larger mines.

ENVIRONMENTAL REGULATION

This section describes the regulatory framework governing hardrock mining on federal lands. The description is not intended to summarize every regulatory program potentially applicable to any mining operation but rather to identify the roles currently played by the major statutory and regulatory programs. It should be noted that for a major hardrock mining operation, 30 or more federal, state, and local regulatory programs may apply. (Appendix C lists the permits required for many recent large-scale mining operations on federal lands in various states.) Small-scale exploration or mining operations may be subject to a more limited array of regulations and programs.

The precise scope of regulation depends on the type of operation; the size of the operation; the types of land, water, and biological resources affected; the state in which the operation is located; the existence of state air and water

---

high standards of living and a wide sharing of life's amenities; and (6) enhance the quality of renewable resources and approach the maximum attainable recycling of depletable resources."

pollution control programs operating with federal approval; and the organization of state and local permitting agencies.

Federal Land Management Agency Responsibilities

Federal land management planning requirements. BLM land use plans under the Federal Land Policy and Management Act (FLPMA)[6] and Forest Service plans prepared under the National Forest Management Act[7] establish the parameters within which surface-disturbing activities may occur on BLM or Forest Service lands subject to the interests created by the General Mining Law. These planning processes are not linked to specific mining proposals, but are intended to guide broad agency management decisions about the use of federal lands and the management of resources on the land. These land management plans do not override the interests acquired by the mining claimant under the General Mining Law, but provide a framework for agency consideration and protection of other resources. Preparation of these land management plans must comply with NEPA,[8] which includes requirements for consideration of alternatives, as well as provisions for public scoping, review, and comment.

BLM and Forest Service regulations. Proposed mining activities on federal lands trigger the application of BLM's 43 CFR Part 3809 regulations (promulgated in 1980) or the Forest Service's 36 CFR Part 228 regulations (promulgated in 1974). The regulations themselves prescribe the review procedures requiring the submission of either notices or plans of operations. The BLM regulations require no notice or other submission for "casual use" operations involving "negligible" surface disturbance; whereas reclamation is required, no standards are set forth and no financial assurance guaranteeing reclamation is required. The BLM regulations require submission of a notice for other operations disturbing 5 or fewer acres, but the BLM does not approve the notice or specify particular operating methods. Reclamation of these "notice-level" operations is required, but no financial assurance is required. Operations disturbing more than 5 acres must submit a "plan of operations" for review and approval by BLM and must post financial assurance in an amount determined by BLM to guarantee reclamation.

The Forest Service requires operations of any size to submit a notice and to submit a plan of operations if the Forest Service determines there will be

---

[6] 43 U.S.C. §§1701-1784.

[7] 16 U.S.C. §§1600-1687.

[8] 42 U.S.C. §4321 et seq.

any "significant" disturbance of the surface. The Forest Service generally requires plans of operations for all mechanized mining or exploration operations on Forest Service lands regardless of acreage. Posting of financial assurance in an amount determined by the Forest Service is required. The Part 3809 and Part 228 regulations establish performance standards intended to assure compliance with, respectively, the FLPMA prohibition of "unnecessary or undue degradation of public lands" and the Forest Service regulatory requirement to "minimize adverse environmental impacts on national forest surface resources," based on the 1897 Organic Act.

The National Environmental Policy Act. NEPA serves to integrate BLM and Forest Service decision making on particular mining proposals with evaluation of other environmental concerns, as well as with other state and federal permitting requirements. The relevant land management agency (BLM or Forest Service) prepares the NEPA documents that inform the manager's decision on the proposed mining operation. Where more than one agency is involved, they reach agreement as to which will be the lead agency or share the lead. Other agencies may become "cooperating agencies" or may elect simply to comment on draft documents in the same manner as the public. The Council on Environmental Quality's regulations encourage federal, state, and tribal agencies that have decision-making responsibilities with respect to a proposed operation—including permitting responsibilities—to become cooperating agencies in preparing the NEPA documents and in performing the required analysis.[9]

For large-scale mining operations on federal lands, the environmental impact statement (EIS) required under NEPA serves as the "spine" of the federal land manager's decision-making process. The EIS process includes requirements for publicly "scoping" the issues and identifying alternatives to be evaluated. It provides for preparation of a draft EIS evaluating alternatives and identifying impacts. The draft is subject to public comment and to review and comment by other agencies. A final EIS must respond to all substantive public and agency comments and serve to inform the decision maker. The final decision is made by a record of decision (ROD), which determines the content of the plan of operations and mitigation requirements.

Smaller mines often trigger only an environmental assessment (EA), a NEPA document intended to assist the federal land management agency in deciding whether environmental impacts are significant. If the EA shows that

---

[9] 40 CFR §1501.6. "Cooperating agency means any Federal agency other than a lead agency which has jurisdiction by law or special expertise with respect to any environmental impact involved in a proposal . . . A State or local agency of similar qualifications or, when the effects are on a reservation, an Indian Tribe, may by agreement with the lead agency become a cooperating agency." 40 CFR §1508.5.

they are, an EIS must be prepared. If not, the agency prepares a finding of no significant impact (FONSI). Unlike EISs, EA/FONSIs are not necessarily subject to public review and comment, although Council on Environmental Quality regulations state that such review is to be afforded in cases where the proposed action "is, or is closely similar to, one which normally requires the preparation of an environmental impact statement" or if the action is "one without precedent."[10] Certain routine activities, such as minor modifications to approved plans of operations, are categorically excluded from NEPA review by BLM or Forest Service procedures. BLM also has a practice of accepting notices for mining operations disturbing less than 5 acres without conducting any environmental review under NEPA.[11] In contrast, the Forest Service prepares EAs in connection with plans of operations for such small operations.

Three public land states have their own "state NEPA" laws that require state agencies to prepare EAs or EISs in connection with activities such as mining. California, Montana, and Washington usually integrate these state processes with the federal NEPA process to avoid duplication of effort. These states have effectively accepted "cooperating agency" or co-lead agency status for NEPA implementation with respect to the review of mining operations. States without "state NEPA" laws must make decisions about their level of reliance on the federal EIS and about becoming a cooperating agency under NEPA on a case-by-case, agency-by-agency basis.

Endangered species, historic preservation, Indian trust responsibilities, and other laws. Other federal laws create consultation and other substantive obligations. The Endangered Species Act (ESA) requires the federal land management agency to consult with the U.S. Fish and Wildlife Service or the National Marine Fisheries Service or both when threatened or endangered species may be adversely affected by a proposed operation.[12] ESA and NEPA processes are often integrated; indeed, the presence or potential presence of listed species may lead an agency to prepare an EIS rather than an EA. The National Historic Preservation Act[13] and the American Indian Religious Freedom Act,[14] also require BLM and the Forest Service to consult with other agencies and with tribes and to consider and mitigate impacts on protected

---

[10] 40 CFR §1501.4(e)(2).

[11] This practice was upheld by the U.S. Court of Appeals for the Ninth Circuit, which held that "BLM does not sufficiently involve itself in the approval process to render notice [level] mine review a major federal action requiring NEPA compliance." Sierra Club v. Penfold, 857 F.2d 1307 (9th Cir. 1988).

[12] 16 U.S.C. §1536.

[13] 16 U.S.C. §470 et seq.

[14] 42 U.S.C. §1996.

resources or interests. The consultation processes are usually integrated with NEPA review. The federal government's trust responsibility to Indian tribes also applies to federal decision making concerning the occurrence of hardrock mining. The trust responsibility stems from a variety of sources, including treaties and statutes. Its applicability is not limited solely to mining operations that occur on Indian lands; it applies also to off-reservation activities on the public lands that affect Indian treaty rights, cultural resources, or other interests.

Coordination with state decision makers. Whereas the decision making outlined above is under the jurisdiction of the land management agencies, memoranda of understanding (MOUs) between the BLM and state agencies and the Forest Service and state agencies establish the links between state environmental regulatory requirements and federal land manager decisions under 3809 and 228 regulations. The Committee reviewed a broad array of these MOUs, which varied significantly in vintage, detail, and specificity. Of the western states, only Arizona lacks a formal interagency agreement related to hardrock mining on federal lands.

Some MOUs involve a state and a single federal land management agency; others are signed by representatives of the state, BLM, and the Forest Service. Some MOUs address only a single issue, whereas others deal with a broad range of management considerations. MOUs frequently indicate whether the federal agency will defer to state decisions, make independent decisions, or share decision-making authority, as well as define how inspection and enforcement, monitoring, and financial assurance will be handled. The MOUs are important, as they seek to avoid duplication while recognizing the sometimes divergent interests of federal and state regulators.

Effect of patenting. For the purposes of regulation, patented lands (or lands acquired by mining operations from the government through exchanges) are treated like any other privately owned lands. They are regulated under the state and federal environmental permitting programs discussed below, but are not specifically covered by the federal requirements discussed above. Specifically, the plan of operations requirements and NEPA compliance associated with approval of plans of operations do not apply to private lands. Nor does the "consultation" requirement under the Endangered Species Act apply to mining on patented lands (unless triggered by some other federal action such as issuance of a permit to fill wetlands); although the prohibition on "taking" of listed species applies whether an operation is on privately owned or publicly owned lands.

Nevertheless, if patented lands are intermingled with unpatented federal lands, the plan of operations and associated NEPA and ESA requirements will apply to the operation. For this reason, some mining operations that include only a small area of unpatented mining claims seek to exchange other lands for these lands, or to obtain patents, in order to simplify the review process and eliminate federal land management oversight.

Environmental Permitting Requirements

In addition to the 3809 and 228 regulations, various other state and federal laws establish environmental requirements applicable to mining operations on federal lands.[15] State environmental regulations of general applicability apply on federal lands and are not preempted by the General Mining Law or other federal laws.[16] The following description highlights the most significant of these environmental laws as they relate to surface management issues.

State reclamation laws. Laws requiring reclamation of hardrock mine sites now exist in all western public land states, although permitting of all operations under these laws has not been entirely completed in Arizona and New Mexico, among the more recent states to adopt such laws.[17] These reclamation laws often establish:

- thresholds for the regulation of exploration operations, small mining operations, large mining operations, and "special" or "designated" mining operations using certain chemical beneficiation techniques;
- application content, review and approval procedures, public participation requirements;
- requirements relating to the characterization of overburden and ores, prediction of acid drainage, and management of acid-generating materials;
- requirements relating to the management of introduced substances related to beneficiation of ores (cyanides, lixiviants);
- requirements for stabilization and reclamation of the site;
- requirements for closure of tailings disposal areas, spent ore areas;
- requirements for revegetation;
- financial assurance requirements, including what the assurance must cover, any minimum or maximum amounts required, forms of

---

[15] See Table C-1, Appendix C, for a list of compilations of state regulatory permitting references consulted by the Committee to assist its primary research on statutory and regulatory requirements.

[16] California Coastal Comm'n v. Granite Rock Co., 480 U.S. 572 (1987).

[17] Ariz. Rev. Stat. §37-932 provides that for mines on federal lands "an approved federal reclamation plan and a financial assurance mechanism . . . consistent with" the state's reclamation law "supersede" state requirements.

acceptable assurance (bonds, letters of credit, corporate guarantee), procedures to establish or revise required amounts, and procedures for release;

- reporting requirements; and
- monitoring requirements.

Small mining operations often have fewer obligations and requirements under these laws (see Table 2-1).

State reclamation laws resemble one another in broad terms, but differ in specificity, approach (technically prescriptive standards versus performance-based standards), and financial assurance requirements. They are generally administered by agencies at the state level organized to regulate mining. In California, however, reclamation is regulated at the county level, subject to oversight by the State Mining and Geology Board.

Surface water quality protection laws. These laws, in general, correspond to the federal Clean Water Act permitting programs. Section 402 of the Clean Water Act requires permits for all discharges from "point sources" to the waters of the United States.[18] A state may operate the permitting program in place of the Environmental Protection Agency (EPA) if the EPA has approved the state's regulatory program. All the western federal land states except Alaska, Arizona, Idaho, and New Mexico operate permitting programs pursuant to EPA delegations. EPA is responsible for surface water discharge permitting in these four states. The permit limits are a combination of technology-based effluent limits and water quality-based limits (where the technology-based limits are not sufficient to meet water quality standards in the receiving waters). Such limits may face further tightening in the future as states conduct assessment of their impaired waters under section 303(d) of the Clean Water Act. Under this section, the states must identify water bodies that do not meet water quality standards with current permit limits, and must establish total maximum daily loads (TMDLs) in order to meet those standards. TMDLs allocate pollutant loads among the sources of discharge.

Section 402 permits typically include numerical standards for the active mining phase and may include different or best professional judgment standards for post-mining discharges. Some states require a "zero discharge" permit for mine facilities designed not to discharge, where there may be adverse effects if discharge occurs. A number of states rely on their water

---

[18] 33 U.S.C. §1342.

TABLE 2-1 Reclamation and Financial Assurance Requirements for Small Mining and Exploration Operations

Federal Requirements

BLM
No notice and no financial assurance is required if "casual use"; operator must prevent unnecessary or undue degradation. Filing of notice is required for all other operations under 5 acres; reclamation is required but no financial assurance.

Forest Service
Plan of operations is required if there is significant disturbance of surface resources; reclamation and financial assurance are required.

State Requirements

Alaska
Under 5 acres requires annual application, letter of intent, annual reclamation statement; no financial assurance.

Arizona
Exploration and mining less than 5 acres are not regulated under reclamation law; no financial assurance. Notice to Department of Water Resources required for drill holes 100 ft. or more deep. Aquifer protection permit can apply to small operations, and requires demonstration of financial capacity; exempts overburden returned to excavation area and materials not chemically leached or that will not otherwise leach concentrations violating aquifer standards.

California
Not regulated under reclamation law if under 1 acre and overburden disturbed is less than 1,000 cu. yd.; no financial assurance. Waste discharge regulations (chemical processes, waste units design and closure requirements) have no size threshold and financial assurance is required.

Colorado
Prospecting not regulated if disturbed area is less than 1,600 sq. ft. and not exceeding two such disturbances per acre, or 5 acres statewide in 24-month period; no financial assurance. Larger prospecting operations require notice of intent and financial assurance of $2,000/acre. Limited impact mining

TABLE 2-1, Continued

operations affecting less than 10 acres and less than 70,000 tons of material have simpler permitting requirements than larger operations except for designated mining operations; financial assurance is required.

Idaho

Hardrock exploration operations must file a notice and reclaim; financial assurance is not required if less than 5 acres. Mines of any size must have reclamation plan and financial assurance. Placer exploration operations must file a notice and reclaim, but financial assurance is not required if less than 2 acres. Placer mines more than 2 acres require reclamation plan and financial assurance limited to $1,800/acre.

Montana

Exploration requires statewide license; license requires bonding of $200-$2,500/acre. Small miners exclusion statement allows an operator to disturb up to 5 acres without a permit; two such sites if at least 1 mile apart. Hardrock small miners are not required to reclaim or post bond; placer small miners must reclaim and post bond not to exceed $10,000. Small operators using cyanide are required to obtain a permit and post bond for that portion of the operation.

Nevada

No reclamation permit is required for exploration or mining operation disturbing and not reclaiming areas of less than 5 acres within 1 mile radius in any calendar year (and not processing more than 36,500 tons annually). No financial assurance is required. Exploration boreholes must be plugged in accordance with regulations of state engineer. Water pollution discharge program applies with no size threshold.

New Mexico

Operations eligible for a general reclamation permit are wet operations (2-100 cu. yd. of material moved per year) or dry operations (up to 200 cu. yd. per year); no financial assurance required. Minimal impact exploration operations (less than 5 acres and no significant impact on water or groundwater or certain other factors) require a permit but no financial assurance. Minimal impact mining operations (less

TABLE 2-1, Continued

	than 10 acres and not otherwise excluded) require a permit, but do not require financial assurance if 2 acres or less.
Oregon	No reclamation permit for mining operations disturbing less than 1 acre of land, or that move less than 50 cu. yd. in streambeds or less than 5,000 cu. yd. otherwise; and no financial assurance required. A special permit is required for exploration that disturbs more than 1 cumulative acre or requires drilling of holes 50 feet or more in depth. Blanket bonding is authorized for exploration.
South Dakota	Exploration is conducted under a notice of intent. A notice is not required for activities that cause "very little or no surface disturbance." Reclamation of exploration operations conducted under notice is mandatory; financial assurance must be posted either based on site-specific conditions or a $20,000 surety for statewide exploration. Small-scale mining operations, defined as affecting less than 10 acres exclusive of access roads and extracting less than 25,000 tons of ore or overburden per calendar year and not using cyanide or other chemical or biological leaching processes; financial assurance not exceeding $2,500 total.
Utah	Notice of intent for exploration and mining operations of less than 5 acres; no financial assurance required; reclamation required in accordance with rules.
Washington	Reclamation law does not regulate mining operations disturbing and not reclaiming less than 3 acres, and exploration operations disturbing less than 1 acre in an 8-acre area; no financial assurance is required for operations below these thresholds. Metal mining and milling act does not have a size threshold.
Wyoming	Requires a notice for exploration by drilling and a license to explore by dozing; the license cannot cover more than 40 acres of disturbance in 4 contiguous 1/16th sections. Both the notice and license require a minimum financial assurance of $10,000. A letter of authorization with no financial assurance

TABLE 2-1, Continued

is given for small disturbances, such as less than 3-inch suction dredging. A small mine permit and financial assurance are required for operations moving 10,000 cu. yd. or less and 10 or fewer acres disturbed in any year.

NOTE: This table addresses primarily reclamation and financial assurance requirements and does not represent all regulation of small operations. For example, small mine operations may require water pollution discharge permits, air pollution permits, wildlife agency approvals, endangered species and cultural resources reviews, or other permits.

pollution control permitting programs to regulate hardrock mining facilities and processes that introduce cyanides, acids, or other substances to the mine site. For example, Nevada's water pollution control permitting process is the primary regulatory framework for control of beneficiation and process units.

In addition to regular section 402 permits for point source discharges from mining operations, the Clean Water Act also requires control of industrial stormwater discharges that enter surface waters, including seepage from leach heaps and waste rock dumps. These are generally regulated pursuant to a general permit or permit-by-rule.[19] General permits under section 402 are also used to regulate water quality concerns associated with placer mining in Alaska.

State groundwater quality laws. Many states have laws designed to protect the quality of groundwaters from discharges from mining and other operations. These laws range from the detailed permitting program in Arizona (requiring aquifer protection permits based on detailed prescriptive standards) to the California mining waste requirements under the Porter-Cologne Act (establishing waste characterization standards and presumptive design standards for management units) to Colorado's groundwater law (which requires each regulatory agency to integrate groundwater protection into its permitting and regulatory decisions).

State water rights laws. State laws in the West subject the withdrawal and use of surface water and groundwater to regulation by the state engineer's office or its equivalent agency. Such review is generally intended to protect beneficial users and assure the integrity of water rights. Thus, mining processes that result in consumptive uses of water or dewatering of aquifers

---

[19] 33 U.S.C. §1342(p).

and surface waters must obtain permits from the state engineer. Water quantity laws also can have some impact on mining practices and approaches, including effects on the post-mining hydrology that must be achieved.

Fish and wildlife laws. Many states have enforceable requirements that relate to fish and wildlife. These may include provisions requiring stream diversion permits (as in California), netting of process ponds to prevent death or injury to migratory birds and terrestrial animals (as with Nevada's industrial pond permitting program), limitations on the timing and location of suction dredging in streams used by anadromous fish (as in Washington), as well as requirements related to definition of post-mining land use objectives and revegetation standards.

Air quality permitting. Air pollution permits are required for most hardrock mining operations beyond the smallest operations. Generally, this is state permitting pursuant to the federal Clean Air Act; it may include controls for fugitive dust, particulates, sulfur dioxide emissions, certain metals, and volatile organic compounds.[20]

Clean Water Act section 404 permitting. A federal permit from the U.S. Army Corps of Engineers is required when a mining activity will result in filling waters of the United States, including wetlands.[21] In some cases, a "general permit" may be available for activities disturbing small areas that are believed to cause minimal impacts. The 404 program requires avoidance of wetlands, minimization of unavoidable impacts, and mitigation of impacts (including compensatory mitigation). Section 404 permitting also triggers a requirement under section 401 that the state in which the activity occurs must certify that the activity will not impair water quality below state standards; states may impose conditions or requirements in connection with their grant of certification.[22]

Other regulation. Other types of regulation typical at hardrock mines on federal lands include Resource Conservation and Recovery Act (RCRA) regulation of certain hazardous wastes and laboratory chemicals as well as certain processing wastes not exempted by RCRA or state laws more stringent than RCRA. Wastes from the extraction and benefication of mineral ores are exempt from hazardous waste regulation under RCRA, as are specified mineral processing wastes, but states may apply more stringent definitions. State solid waste regulations also apply to various wastes, including garbage, construction debris, and other wastes not exempt under state law. Other regulatory programs include the Safe Drinking Water Act regulation of underground injection and protection of source waters; protection of cultural resources and

---

[20] 42 U.S.C. §§7401 et seq.

[21] 33 U.S.C. §1344.

[22] 33 U.S.C. §1341.

archaeological permitting; reporting requirements under the Emergency Planning and Community Right-to-Know Act; spill reporting obligations under the Comprehensive Environmental Response, Compensation, and Liability Act, and liability under that Act for releases of hazardous substances to the environment (including liability for natural resource damages); Uranium Mill Tailings Radiation Control Act regulation of uranium tailings disposal sites; local regulation, zoning, and occupancy requirements; transportation regulation (including road closures, relocations, crossings); and other requirements.

Section 313 of the Emergency Planning and Community Right-to-Know Act requires certain types of operations annually to report any releases to the environment, accidental or planned, of the 650 listed chemicals (including metals and other substances identified as hazardous), as well as amounts of chemicals involved in certain other types of activities, such as recycling or destruction. Originally implemented in 1987, the Toxic Release Inventory regulations under this section were revised in 1997 to require hardrock mining operations to complete this reporting. The first report from hardrock mines was due on July 1, 1999, for 1998 calendar year releases.

Regulation of uranium industry. The regulatory structure that governs the uranium industry is unique because of the role of the Nuclear Regulatory Commission, which regulates all aspects of the uranium industry that involve processing of uranium, including the *in situ* solution mining of uranium. The EPA, or a state with a delegated program, also regulates the *in situ* solution mining of uranium because the injection wells are considered to be Class 3 wells under the Underground Injection Control Program (40 CFR 140-143). State groundwater protection and water rights laws also apply. Traditional mining of uranium is regulated in the usual fashion by other federal and state agencies. A more detailed discussion of the uranium industry is in Appendix A.

RELATIONSHIPS AMONG STATE AND FEDERAL ENVIRONMENTAL REGULATIONS

Crosscutting issues under all the federal, state, and local programs include permitting, monitoring and reporting requirements, inspection and enforcement, reclamation, surety, bond release, and public participation opportunities. Some of these relationships are defined by MOUs, while others are worked out more informally. The current array of MOUs differ significantly in their specificity.

Given the variation in topography, climate, and area of federal lands open to hardrock mining in any state, differences in state laws, and local differences in public attitudes toward mining, consistency among state MOUs may not be necessary or even desirable. Testimony provided by federal and state officials at meetings of the Committee indicated that the MOUs generally serve their stated purposes. However, of more relevance is whether federal and state laws each need to cover all topics of concern, or whether coordination of the respective laws can adequately address hardrock mining impacts. For example, apparent gaps in federal regulatory authority (e.g., groundwater quality) may be filled by state authority. On the other hand, federal reclamation bonding requirements, such as for small operations on Forest Service lands, sometimes fill a state regulatory vacuum. Unless there is strong coordination, gaps or duplication of effort may occur.

How do these varying programs relate to one another? Federal agencies that manage lands open to mineral exploration and development regulate hardrock mining and exploration by managing the land and regulating the land's uses. Congress, mostly acting through laws administered by the EPA, has assigned permitting and regulation of water quality and air quality to the states (except where states have elected not to seek authorization and have left permitting to EPA). Most states have taken on these responsibilities. Water quantity has historically been subject to state regulation. Groundwater quality (in the absence of federal legislation) has also been largely a matter of state regulation.

Protection of cultural resources on lands open to mineral extraction is of concern to federal agencies through the National Historic Preservation Act, to states through their state historic preservation offices, and to Indian tribes and others with cultural heritage in the mining areas, such as residents of historic mining towns. This dispersion of management and concerns for environmental and cultural protection among federal, state, tribal, and local entities raises the issue of which entity has, or should have, the leading role in managing and regulating activities that may create environmental degradation.

Virtually all areas of concern about environmental degradation or protection of cultural resources are addressed in some fashion by regulation and permitting procedures; however, there are times when higher levels of environmental or cultural protection may be desired for site-specific reasons. These resource protection issues can often be addressed through federal resource management planning processes, which allow input from all interested parties.

In general, the state and federal environmental permitting agencies are the first-line enforcers. The BLM and Forest Service may also take enforcement action under their respective 3809 and 228 regulations when the activity results in violation of those regulations or violates the terms of the plan of operations or the notice requirements.

State regulatory agencies, the EPA, the U.S. Army Corps of Engineers, or the U.S. Fish and Wildlife Service will ordinarily take the lead in enforcement when a regulatory standard has been contravened—such as violation of an effluent guideline, failure to report a release, or causing the death of migratory birds or harm to endangered species. Federal land managers may act independently (even where a regulatory agency has also acted) to assure that land management objectives not necessarily reflected in an abatement order issued by another agency are satisfied.

In general, the existence of multiple regulatory programs helps to assure that at least large-scale mining on federal lands is subject to substantial scrutiny. At the same time, however, the complexity of the various programs can make the system difficult to understand, approach, and implement efficiently. As a result, mining regulation, permitting, monitoring, reclamation, closure, and post-closure becomes a series of negotiations carried on against a background of regulatory requirements and programs. This means that governmental regulators at all levels need a significant degree of sophistication and training in order to make these programs efficient and effective. The programs do not—and cannot—operate in cookbook fashion. Thus, many of the findings and conclusions of this report involve mixed issues of regulatory adequacy and implementation.

TIME REQUIREMENTS FOR ENVIRONMENTAL REVIEW AND PERMITTING

Examination of timing provisions and requirements in state and federal laws and regulations reveals a variety of approaches by the agencies. The amount of time required for project review and permitting is influenced as much by implementation practices of state and federal agencies as by statutory and regulatory requirements. For example, while NEPA comment and review requirements under the Council on Environmental Quality's regulations[23] could in theory allow a project to proceed from notice of intent to prepare an EIS to a record of decision in approximately six months, the committee found that large-scale mines on federal lands require between 18 months and 8 years to complete both the EIS review and all the permitting and other approvals by state and federal agencies with jurisdiction over the mining operations. NEPA regulations do provide some opportunities to address timing issues. The regulations expressly encourage federal agencies to set time limits appropriate to individual actions. They also provide that an agency "shall set time limits if an applicant for the proposed action requests them."[24]

---

[23] 40 CFR §1500.
[24] 40 CFR §1501.8(a).

Some laws have explicit timing requirements. For example, Colorado law provides that a decision must be made on the state reclamation permit within 120 days after receipt of the application, unless the operation is particularly complex or other circumstances require more time. The law provides for an additional 60-day review period under such circumstances. To afford sufficient time for review, Nevada's water pollution regulations state that applications must be submitted at least 165 days before the intended commencement of operations; the regulations also require a pre-application meeting with the regulatory agency to identify data needs and avoid unnecessary delays in determining that permit applications are complete. Nevada's reclamation regulations tie the timing of the reclamation permit issuance to completion of the federal EIS process. Some states, such as California, have adopted time requirements for their "state NEPA" processes. However, even where state laws impose time limits, operators noted that there can be implicit pressure to agree to extensions of time beyond these limits in order to avoid risking a denial based on lack of information desired by an agency.

The time required for environmental review and permit issuance is also related to the completeness and technical adequacy of the permit information provided by the operator. It can take a significant time for the applicant to acquire an adequate data base. Timing of environmental review and permitting is affected by agencies' ability to coordinate with one another, as well as by the availability of sufficient agency staff and technical resources. Where coordination among state and federal regulatory agencies is high, environmental review and permitting appears to be faster—at least in areas that are not highly controversial due to publicly competing interests and values. Where separate agencies engage in serial permitting, rather than coordinating their review efforts, the process—including data gathering—can take longer. These implementation relationships are explored further in Chapter 3.

3

Adequacy of Environmental Protection

> *As specified by Congress, the study shall identify and consider . . . the adequacy of federal and state environmental, reclamation, and permitting statutes and regulations applicable in any state or states where mining or exploration of locatable minerals on federal lands is occurring, to prevent unnecessary or undue degradation.*

This chapter responds to Congress's second request to the Committee, and includes assessment of statutes and regulations, implementation of the regulations, and related matters to protect the environment. The Committee's assessments are presented as a series of issues related to the adequacy of environmental protection and regulatory efficiency related to mining. Apart from the issues that are identified, the Committee noted the significant improvements in mining-related environmental protection in recent decades. The Committee also noted that the staffs of federal land management agencies and the federal and the state regulatory agencies are generally dedicated to their tasks. Finally, all stakeholders the Committee encountered appeared to be concerned about environmental protection, even though they may disagree about the means and extent of that protection.

An evaluation of the adequacy of environmental protection related to mining on federal lands should include an analysis of the statutes to determine if they are protective of the environment or if they contain gaps or duplications. A regulatory program, however, is more than a collection of statutes. Laws can differ in their implementation, and their interpretation can be modified by court decisions. Laws are amplified and clarified through regulations. The regulations in turn are supplemented by published guidelines, policy statements, and administrative interpretations, which may not be consistent over time or among offices. The Committee has reviewed several recent compilations of regulatory requirements (see Appendix C).

The implementation of regulations through permits depends on such factors as the training and experience of the permit writers, their workload and staffing levels, their specific knowledge of the proposed operations, and their interpretation of rules and policies. Similar factors may affect the quality and effectiveness of the site inspections and monitoring that determine if mining activities are in compliance with operating permits. Finally, the implementing authorities must have and be willing to use enforcement tools when non-compliance occurs. Ultimately, the adequacy of state and federal regulatory programs for protecting the environment must be measured by what happens on the ground.

In this Chapter the Committee first presents a brief discussion of the essential elements of a regulatory program. The Committee then discussed a number of issues related to the adequacy of environmental protection of hardrock mine sites. The following issues are discussed and assessed: regulatory issues; regulatory implementation issues; scientific issues; the need for early stakeholder consultation; reclamation, closure, and post-closure management; regulatory efficiency issues; and public involvement issues. The issues presented herein lead to the recommendations presented in Chapter 4.

ELEMENTS OF A REGULATORY PROGRAM

The Committee identified the following 11 elements of an effective regulatory program:

(1) *Information* to describe pre-mining conditions, potential changes to the environment resulting from a project, and proposed post-mining/reclamation land use conditions;

(2) *Models and tools* to project and assess the consequences of these changes;

(3) *Standards and criteria* to determine if the changes comply with permit requirements, or are otherwise acceptable;

(4) *Monitoring* to demonstrate that standards and criteria are met.

(5) *Reporting and analysis* of the monitoring results to assess compliance with standards and criteria and trigger appropriate response to non-compliance;

(6) *Appropriate responses* to correct noncompliance and assure it will not reoccur;

(7) *Enforcement* actions to assure adequate responses to non-compliance;
(8) *Financial guarantees* to protect the environment if the operator fails;
(9) *Personnel and resources*, properly trained and supported, to adequately implement and administer the regulatory program;
(10) *A management information system* to track the effectiveness of the regulatory program; and
(11) *Stakeholder communications* to assure that the entire process is open to appropriate public scrutiny.

Some of the elements are procedural (e.g., reporting and enforcement), and others describe the resources needed to make the procedures effective (e.g., models and tools, monitoring, and the management information system). Recognizing in advance the need to satisfy these eleven elements will add to the timeliness and adequacy of the regulatory program. These essential elements are discussed below and were used to identify the issues discussed later in this chapter.

Information

Information needed to describe environmental changes that result from a mining project can be divided into two categories: (1) environmental data and other information describing the site's existing, and, if available, pre-mine conditions; and (2) proposed plans of operations, reclamation plans, post-closure management plans, and other information required to identify the physical activities proposed for the site, and whether mining conditions can support the post-mining land use. Both sets of information must be adequate to determine the type and extent of impacts.

Models and Tools

The models and tools needed to project and assess the consequences of changes in baseline conditions resulting from the activities proposed for the site include air quality emission factors and models, acid-generation prediction models, pit lake water quality models, and hydrological models, among others. Current models and tools have varying degrees of uncertainty and have been subjected to varying degrees of calibration and verification (see Appendix D).

Standards and Criteria

Standards and criteria are required to determine if the physical activities proposed for the site are in compliance with permit requirements. These standards and criteria are derived from the authorities applicable to the regulation of mining. They range from specific numerical standards, such as those included in federal and state Clean Water and Clean Air Act requirements, to the more qualitative requirements, such as those typically contained in state reclamation laws.

Monitoring

Appropriate environmental monitoring is essential to establish the effectiveness of containment measures, demonstrate compliance with standards and criteria, and direct responses to non-compliance (see Appendix B for discussion of long-term monitoring). Care should be taken to assure that monitoring requirements are adequately comprehensive and include the necessary quality control measures and documentation to be credible to the public. To the extent feasible, monitoring programs should include "early warning" elements to detect potential non-compliance far enough in advance to allow a suitable response that will prevent violation of standards and criteria.

Reporting and Analysis

Adequate reporting of monitoring results is required to assess compliance with standards and criteria and to trigger appropriate responses to incidents of non-compliance or, in the case of early warning monitoring, incidents of potential non-compliance. Reports should include tabulations, analyses, interpretations, and summaries that will allow the public to understand the results, will clearly support the conclusions, and will target actions to correct non-compliance.

Corrective Actions

Mine operators should quickly respond to correct actual or potential environmental problems identified through inspection and monitoring activities. These responses should be designed and engineered by qualified professionals, and if non-compliance has occurred, should be reviewed and approved by the agencies with jurisdiction. Implementation should be monitored and documented with the same care given to original construction.

Enforcement

Enforcement tools are required to assure adequate responses to violations and to assure that all of the required program elements are implemented. However, enforcement tools are preferably a seldom used component of an effective regulatory program. It is important to distinguish between circumstances presenting imminent danger to human health or the environment and other occurrences of non-compliance. Any enforcement mechanism should include appropriate due process for the violator and should emphasize compliance rather than operator penalty.

Financial Guarantees

Reclamation financial guarantees have long been an element of both state and federal programs, and are needed to assure that mines are reclaimed, even when the operator fails or is financially unable to undertake reclamation. These costs should not become a public burden. More recently, financial assurances are being required for predicted or actual environmental impacts during or after mining if the operator fails to appropriately correct identified problems or does not respond to pollution releases. In addition, most jurisdictions are now examining the need to assure post-closure management of reclaimed mine sites. The various financial mechanisms should be secure and sufficiently liquid to allow responses to near-term needs.

Personnel and Resources

Adequately trained personnel and supporting resources are the key to effective implementation of a regulatory program. Adequate training is needed to maintain skills in the face of changing technology and to maintain an understanding of mining and environmental requirements. Acquiring and maintaining the necessary skills requires adequate funding and other technical resources. To complete inspections in remote areas, the federal land management agencies should consider the use of helicopters. Although costly, this option may be especially useful when notices of intent are filed in remote areas and BLM has only 15 days to respond.

Management Information System

A management information system (MIS) is essential to regulatory programs. Such a system tracks the effectiveness of the regulatory effort and allocates the necessary personnel and resources to maintain adequate environmental protection and to continuously improve overall program effectiveness. An MIS can provide the basis for effective stakeholder communications, as well as the number and types of mines; summaries of monitoring, inspection, and enforcement information; and the status of reclamation, financial assurances, and outstanding violations.

Stakeholder Communications

Effective, two-way stakeholder communications are essential to assure that the entire regulatory program is appropriately open to public scrutiny. An open and well-communicated regulatory program is essential in building public confidence that effectiveness has been achieved and that federal lands are being adequately protected.

REGULATORY ISSUES

The Committee did not have sufficient information to evaluate fully the environmental impacts of modern hardrock mining. Regulation of mining will limit and control many of these impacts, but mining will still alter landscapes and environmental resources because regulations generally are not designed to prevent all impacts, because some impacts are not addressed by regulations, and because it is unreasonable to expect there will not be violations or failures of the regulations.

While most of the current environmental impacts from hardrock mining are related to the over 200,000 inactive and abandoned mines (EPA, 1997b), some ongoing mining operations also affect environmental resources. (Appendix B contains a summary of the kinds of potential environmental impacts that can occur at hardrock mines on federal lands.) Nearly 20% of the mining sites inspected by EPA and the states between August 1990 and August 1995 were subject to enforcement actions. Nearly all the violations involved either the Clean Water Act, the Clean Air Act, or the Resource Conservation and Recovery Act (EPA, 1997b). This suggests that some hardrock mining operations cause environmental impacts, even under current regulatory conditions, and that these impacts are subject to regulatory enforcement actions. The Committee has no information regarding the effectiveness of these enforcement actions.

The Committee reviewed several assessments of the substance and adequacy of federal and state laws and regulations pertaining to hardrock mining throughout the western United States. The Committee heard testimony from speakers who contended that a compelling case had not been made to revise the BLM regulations and from other speakers who contended that the regulations do not adequately protect the environment. The Committee identified issues or gaps that reflect inadequacies in the statutes and regulations related to environmental protection and regulatory efficiency. In spite of these issues or gaps, and based on numerous discussions and meetings with federal and state regulators, mining industry representatives, and the public, the Committee finds that the existing regulations are generally well coordinated and effective. These issues or gaps related to regulation are discussed below.

Role Conflicts in Land Management Agencies

The Committee found that federal land management agencies have potentially conflicting roles. First, they manage the land for the multiple, and often conflicting, uses provided for by laws and regulations. In a sense, they are hosts for miners, lumbermen, and recreation visitors. Second, they are regulators responsible for ensuring that all activities conform to laws and regulations and that the environment is protected. Third, they are responsible for implementing the Mining Law of 1872, which appears to assign a priority to hardrock mining on the lands they manage. This leads to the following questions: Can they fairly and evenly discharge their historic role as host to land users and accommodate their growing role as regulators? Can they balance the interests of miners and the general public by issuing timely permits that reasonably protect the environment? Can they effectively implement the regulatory authorities that statutes provide?

Other agencies face similar conflicts. For instance, the Environmental Protection Agency (EPA) is responsible for evaluating the adequacy of environmental impact statements (EISs). Yet, the EPA should be a cooperating agency in the NEPA process for preparation of EISs. In some instances, when conflicts between management and regulation apparently became too serious, Congress assigned different roles to different agencies, sometimes by dividing the original agency into independent parts. One partial solution to this problem is to provide easy public access to agency information and activities so the public could perform an oversight function. Recommendations 10, 11, and 13 address the issue of interagency cooperation and public participation. The

Committee believes the federal lands can be managed efficiently and effectively despite these potential conflicts.

Regulatory Standards

There are two types of regulatory standards used to limit environmental effects of mining: performance-based standards and technically prescriptive standards. Performance-based standards, sometimes referred to as outcome-based standards, specify the desired result or outcome rather than a method, process, or technology. The BLM's standard to avoid "unnecessary or undue degradation" and the Forest Service's standard to "minimize adverse environmental impacts" are both performance-based standards. These general standards articulate the agencies' overall management objectives, and can be used as the foundations or goals on which more specific standards are established. Technically prescriptive standards sometimes referred to as design standards, prescribe the specific techniques to be used to achieve adequate environmental protection. Standards that specify the type and thickness of liners below waste dumps and tailings piles are examples of technically prescriptive standards.

A strength of performance-based standards is that they allow maximum consideration of the particular site-specific conditions and technology options in deciding how to achieve environmental goals. They also encourage the development of new and more cost-effective approaches to achieving these goals. Some disadvantages are that they may require regulators to have a greater degree of technical sophistication to assess the ability of proposed methods to meet the standards. Also, mine operators may have less initial certainty as to the methods that agencies will deem acceptable in meeting performance standards; the public may be confused by the approval of different methods to meet the standards from site to site. These disadvantages may, however, be addressed through training and development of clear guidance manuals and procedures. Examples of Canadian guidance materials that are used routinely by mine operators in Canada and the United States include Price and Errington (1998) and Price (1997). Manuals such as these that provide technical guidelines but acknowledge site-specific variability could be created for issues related to environmental protection at mine sites.

A strength of technically prescriptive standards is that they minimize ambiguity and uncertainty about requirements of the regulated facility. Technically prescriptive standards also have disadvantages. For example, they typically assume that a national or statewide standard is appropriate for all locations and conditions and do not allow for site-specific conditions that may warrant special considerations. They also are insensitive to technological change and may inhibit improvements. Recommendation 9 in Chapter 4 reflects the Committee preference that BLM and the Forest Service rely predominantly on performance-based standards.

Financial Assurance

The requirement for financial assurance for mining activities on federal lands has been inconsistent, depending in part on the type and size of the activity and the state where it occurs. For example, the Committee observed instances of recently abandoned but unreclaimed exploration and mining sites that had not been covered by any financial assurance. Financial assurance for such disturbances beyond casual use could discourage unreclaimed abandonment and provide funds for reclamation.

The Committee also found that long-term water treatment and monitoring at mine sites generally does not carry financial assurance at either the state or federal level. A few states have adopted regulatory programs that authorize agencies to require financial assurance for long-term water quality protection. The BLM and the Forest Service have negotiated such financial assurances, but on a case-by-case basis for some mines on federal lands.

Under current federal land management regulations all mines with operating plans are required to have some type of financial assurance for reclamation. Notice-level operations under the BLM regulations do not have to provide financial assurance. Notice-level activities can include exploration, mining, and processing operations, provided they disturb less than 5 acres. Casual use activities pose insufficient environmental risk to warrant financial assurance.

Appendix E summarizes the types of available financial assurance. Based on the Committee's findings, inadequate protection of the public and the environment caused by current financial assurance procedures is a gap in the regulatory programs. Recommendations 1 and 14 in Chapter 4 address financial assurance for mining activities.

Modification of Plans of Operations

Several commentors, including federal agency personnel, expressed concern to the Committee that BLM and the Forest Service lack sufficient authority to require prompt modifications of plans of operations when the potential for degradation is identified. Examples warranting revisions could include unexpected acid drainage, problems with water balance, potential inadequacy of containment structures, and discovery of water quantity impacts on wells and springs (see Appendix B). The authority to require operators to make modifications for cause is necessary to prevent unnecessary or undue degradation of federal lands. BLM regulations state that "at any time during operations under an approved plan, the operator on his/her own initiative may modify the plan or the authorized [BLM] officer may request the operator to do so." (43 CFR 3809.1-7). A "significant" modification must be reviewed and approved by BLM "in the same manner as the initial plan," that is, with appropriate NEPA review, among other requirements. If an operator fails to submit a modification requested by the authorized BLM officer "within a reasonable time, usually 30 days," the officer may recommend to the BLM state director that the operator be required to submit a modification.

In deciding whether to require a modification, the state director must determine whether "all reasonable measures" were taken by the authorized officer "at the time the plan was approved to ensure that the proposed operations would not cause unnecessary or undue degradation"; that the disturbance now resulting from operations under the approved plan or from unforeseen circumstances "is or may become of such significance" that modification is "essential" to prevent unnecessary or undue degradation; and that the disturbance can now be minimized using "reasonable means." Operations continue under the prior approved plan until a modified plan has been submitted and approved. However, if the operations are actually causing unnecessary or undue degradation, the state director must advise the operator of "reasonable measures" to be taken pending review and approval of the modifications. The operator must immediately take all necessary steps to implement those measures within a "reasonable period" established by the state director.

Forest Service regulations provide that "at any time during operations under an approved plan of operations, the authorized officer may ask the operator to furnish a proposed modification of the plan detailing the means of minimizing unforeseen significant disturbance of surface resources." If the operator does not furnish a proposed modification within a time deemed reasonable by the authorized officer, the authorized officer may "recommend to his immediate superior that the operator be required to submit a proposed modification of the plan" (36 CFR 228.4[e]). The supervisor must find that all reasonable measures were taken to predict the impacts in reviewing the original plan of operations, that the unforeseen disturbance is significant, and that it can be minimized using reasonable means. "Lacking such determination that unforeseen significant disturbance of surface resources is occurring or probable and that the disturbance can be minimized using reasonable means, no operator shall be required to submit a proposed modification of an approved plan of operations" (36 CFR 228.4[e][3]). As with BLM regulations, the Forest Service regulations provide that operations may continue under the previously approved plan of operations pending approval of a modified plan unless the supervisor determines that measures are needed to prevent unnecessary or unreasonable injury to surface resources.

Disputes have arisen over the ability of the BLM and Forest Service to require modifications in various settings based on the language of the regulations. Such disputes include (1) whether issues of concern were identified in the review of the original plan, or whose fault it was that they were not addressed and (2) issues of the "reasonable means" that may be considered in a modification proceeding. The lack of clarity in these standards and the backward-looking nature of the findings required by the regulations have created uncertainties and inefficiencies in addressing impacts promptly. Recommendation 4 in Chapter 4 addresses the issue of modifying plans of operations to protect public lands.

Different Regulatory Approaches for Exploration, Mining, Extraction, and Mineral Processing

Exploration, mining, and processing present different levels of environmental risks (see Appendixes A and B). Similarly, different types of exploration and different types of mining and processing present different levels of environmental risk. The Committee found that regulatory efficiency does not necessarily reflect the level of environmental risk (e.g., exploration projects creating little potential environmental impact may take an extraordinarily long time to permit).

It appears that the Forest Service and BLM do not adequately tailor regulations and permitting to match a project's potential for environmental damage. The Forest Service requires that financial assurances be provided and environmental assessments and plans of operations be prepared for any activity beyond "casual use." The Forest Service requirement for plans of operations seems restrictive for small exploration projects that have little environmental impact. The BLM, on the other hand, requires neither financial assurance nor plans for activities disturbing less than 5 acres, requiring only that a 15-day review period precede any operations. The BLM approach seems more appropriate for exploration projects, except for the lack of financial assurances for activities on less than 5 acres.

Information developed as exploration proceeds is used in decisions on steps that follow. At each stage of exploration, new findings help the operator decide whether to abandon the project, modify work in the already disturbed area, or extend the program in a particular direction on the ground. Extensions must be approved promptly if work is to continue and particularly if there is a limited exploration season. The Committee heard suggestions that this permitting process should consider the past environmental performance of the operator and a maximum of 5 unreclaimed acres as bases for extending the exploration area.

On the other hand, other mining activities, such as mine development, extraction, and mineral processing, can generate significant environmental impacts, and BLM's policy of requiring neither financial assurances nor plans of operations for such activities that disturb less than 5 acres may not be sufficiently protective of the environment or the government's fiduciary obligations.

The Committee emphasizes the need for regulatory efficiency (e.g., short permit time and permit restrictions that match the potential for environmental damage). Recommendations 2 and 3 in Chapter 4 offer approaches to minimize the inconsistencies in the different regulatory approaches for mining disturbances.

Protecting Valuable Resources and Sensitive Areas

The agencies have at least 5 mechanisms for protecting valuable resources and sensitive areas from mining. The first is to formally withdraw federal land from mining and mineral development in recognition of extraordinary natural, scenic, or cultural values that are of such significance that other competing resource uses must be excluded. The withdrawal authority is also used to protect administrative and other public facilities.

The second mechanism is to prepare land management plans that identify natural and cultural resources and allocate competing uses accordingly. These plans specify limitations and mitigation that apply in particular locations and to particular uses, such as mining. These plans are intended to be based on thorough resource inventories and to be developed with broad public input so that they balance the needs of all constituencies. For BLM lands FLPMA also specifies that land use plans give priority to the designation and protection of areas of critical environmental concern (FLPMA, Sec. 202[c][3]). These are areas recognized as needing special management "to protect and prevent irreparable damage to important historic, cultural, or scenic values, fish and wildlife resources or other natural systems or processes, or to protect life and safety from natural hazards" (FLPMA, Sec. 103). Designated areas of critical environmental concerns (ACEC) are usually withdrawn from hardrock mining.

The third mechanism for protecting valuable resources and sensitive areas is the use of advisory guidelines that identify categories of resources or lands that deserve special consideration. Use of such guidelines is consistent with the principle that regulatory decisions should be based on site-specific evaluations and conditions. For instance, in many areas of the western United States, healthy riparian habitat is scarce and has high value for wildlife or as a buffer to protect stream quality. In these cases, the flexible regulatory framework would suggest that riparian areas should be valued and be provided reasonable protection in site-specific decisions.

A fourth mechanism is through use of BLM's Visual Resource Management System (VRMS) which recognizes that public lands have a variety of visual values (BLM, 1986; BLM/EPA, 1998). The BLM uses the VRMS to evaluate the potential effects of an action on existing visual resources. A VRMS classification is part of BLM's land use planning process. It is important to note that the VRMS classification applies to all activities on BLM lands, not just mining-related activities. The system is also used as a guide to help ensure that every attempt is made to minimize potential visual impacts of a proposed action. The basic philosophy and uses of the system are described in the *Visual Resource Contrast Rating Handbook* (BLM, 1986).

A fifth important mechanism by which BLM and the Forest Service routinely protect environmental resources that are not governed by specific laws is the adoption of site-specific mitigation measures imposed on a plan of operations after the environmental assessment or EIS process. This is the most common way for the federal land management agencies to protect cultural values, riparian habitat, springs, seeps, and ephemeral streams that are not otherwise protected by specific laws.

The Forest Service appears to have ample authority to provide protection. BLM's interpretation of unnecessary or undue degradation, however, does not explicitly state that the BLM has the ability to provide such reasonable protection. It only indicates that the BLM can require reasonable mitigation for damage that does occur. In practice, however, the BLM has protected some resources that are not protected under other statutes. In most of these cases the applicant has either not formally protested the BLM's proposed protection or the Department of the Interior's Board of Land Appeals (IBLA) has upheld the BLM decision.

The right to mine on federal lands under the Mining Law of 1872 is conditioned on validation of discovery, which depends in part on the expected total costs of production, including those for environmental compliance and mitigation. The IBLA has stated this principle as: "First of all, the mere filing of a plan of operations by a holder of a mining claim invests no rights in the claimant to have any plan of operations approved . . . If the costs of compliance render the mineral development of a claim uneconomic, the claim itself is invalid and any plan of operations therefor is properly rejected" (IBLA 97-307, 97-506, 97-510: Great Basin Mine Watch et al. decided November 9, 1998). Recommendation 15 addresses the agencies' authorities to protect valuable resources that are not protected by other laws.

Protecting Cultural Resources and Tribal Interests

Several presentations to the Committee emphasized the difficulty of identifying cultural resources and tribal issues, and the time and cost that can be involved in locating and dealing with these issues. Government agencies have not expended as much effort in identifying, cataloguing, and protecting these resources as they have on behalf of the natural environment.

Cultural resources may include any works of mankind, be it American Indian or immigrant, pre-historic or of the current era. They may also include natural landscape features that are specially significant for cultural reasons. Proposed mining activities commonly will occur in previously mined areas because that is where new mineral deposits are likely to be found. New mining could put at risk historic ghost towns, railroads, bridges and other mining-related relics. These features are part of the settlement of the West and the industrial history of this country, and they often are afforded cultural preservation consideration.

Tribal interests may include, in addition to easily recognized cultural resources such as burial sites, less obvious features or locations such as seasonal campsites, or water sources such as springs. Federal land management agencies have limited capability to identify such areas unless they are assisted by tribal members. Some tribes have only recently begun to identify and protect their cultural and historic interests. Tribes may be reluctant to identify important areas or resources in order to protect their sanctity or prevent intrusion and plunder by others. This reluctance makes it difficult to fully consider tribal interests in a NEPA review and permitting process that relies on full participation of stakeholders and full disclosure and discussion of relevant information.

The Committee learned that efforts to identify cultural and tribal resources can significantly delay the completion of environmental impact studies. Tribes, for instance, may fail to respond or respond late in the NEPA process. In some cases this may be a result of a tribe having limited resources to deal with such issues. However, such delays can disrupt the efforts of other stakeholders to conduct an efficient process. In other cases, all tribes with potential interests in a land use proposal are not contacted, or are not contacted in a timely manner. Similar problems may arise with other types of cultural resources, even though most states have agencies that are responsible for cultural resources.

Federal agencies and many states now employ specialists in cultural and tribal resources. The Committee was told that better notification of land use proposals is now generally given to potentially interested stakeholders. Some tribes enter into confidential disclosure agreements with land management agencies. The Committee is not clear about the extent to which existing laws and regulations, such as the National Historic Preservation Act, the American

Indian Religious Freedom Act, and various state laws, adequately protect cultural resources and tribal interests. It also is not clear if cultural resources and tribal concerns are protected by either the Forest Service regulations providing the agency with authority to "minimize adverse environmental impacts on National Forest resources" or by BLM's authority to avoid "unnecessary or undue degradation." Providing adequate protection to these resources and concerns without unreasonably delaying the permitting process requires a delicate balancing of interests. This can best be achieved by the early participation of all potentially affected interests in the NEPA and permitting process, which is the focus of Recommendation 10 in Chapter 4.

Issues of Small Mining Operations

Several individuals, including both small and large operators, stated that small miners or small operators play an important part in the mining industry, especially in the discovery of new mineral deposits. Although the environmental risk may be small compared to large operations, small operations are sometimes abandoned without adequate reclamation, requiring considerable agency time and imposing financial liabilities on the land management agency.

Recognizing the special circumstances of small operators, BLM regulations exempt operators disturbing less than 5 acres from the requirements of an operating plan or financial assurance (see Chapter 2). The small operations need only submit a "notice" to BLM indicating what they intend to do. BLM must respond to the notice within 15 days if it chooses to restrict or modify the proposed operations. Such notice-level activities require much less effort and time to obtain permission to begin operations. Nevertheless, such operations, particularly if they involve mine development, extraction, or mineral processing activities, should be reviewed, and questions should be raised if the regulatory program is inadequate.

The Forest Service provides no such exemption for small operations, although most small operators need submit no more than a short environmental assessment and operating plan with their applications. Nevertheless, agency staff restrictions often delay the preparation of EAs for several months, thus delaying the decision on the application.

The Committee visited notice-level sites on BLM lands and received testimony from BLM, the Forest Service, state regulatory officials, and industry and concerned citizens regarding the adequacy of current regulations. Degradation on notice level sites included open trenches and unreclaimed roads, abandoned mining and processing equipment, and unreclaimed stockpiles of ore and waste rocks. This is a burden to the agencies, and to the taxpayers, who will pay for cleanup. Regulatory agencies find it difficult to implement procedures that protect and promote the beneficial activities of small mining operations without allowing significant environmental impacts or imposing financial liabilities on the public. Issues related to small mining operations are addressed in Recommendations 1, 2, and 3 in Chapter 4.

Mining Reclamation of Abandoned Mine Sites

During presentations to the Committee, representatives of mining companies and regulatory agencies expressed frustration with current laws that discourage new mining operations from disturbing or reclaiming abandoned mine sites. Although the charge to the Committee did not include consideration of abandoned mines on federal lands, the affect of these historic features on future mining, and their related impacts, requires some mention.

New mines commonly develop at or near the sites of abandoned mines. New mining activities have the potential to reclaim damage from earlier mining activities. However, some federal environmental laws, such as the Clean Water Act, the Comprehensive Environmental Response, Compensation and Liability Act (CERCLA), and potentially the Resource Conservation and Recovery Act (RCRA), discourage these opportunities and encourage companies to design their operations to avoid redisturbing previously disturbed areas. Similarly, operators are discouraged from reclaiming previously mined areas in the vicinity of new mining operations, even though the work would be beneficial and voluntary on the part of the miner. Reclamation of some abandoned mines has been integrated into current operations, and nearby historic waste disposal sites have been reclaimed in connection with ongoing mining activities. This is done, however, at the miners' risk. As a result of existing statutes, mine operators conducting work at an abandoned mine site could incur liabilities. Some states have environmental laws that may pose additional barriers to voluntary clean up. Therefore, operators avoid redisturbance of abandoned mine discharges and arrange their own discharges so that the pre-existing discharge is not affected. This is not in the best interests of environmental protection; opportunities for remediation are missed and new operations are, in effect, encouraged to disturb undisturbed lands. Recommendation 7 in Chapter 4 addresses the reclamation of abandoned mine sites.

Temporary Mine Closures

Changes in commodity prices and other factors can influence the economics of mining and can result in an operator temporarily closing a facility. The duration of the closure is generally not known in advance, hence steps to protect the environment under these interim conditions may not be taken. In some cases concern was expressed that operators were using this temporary status to delay or avoid proper closure and cleanup of the facility. Lacking regulations that cover these situations, regulatory agencies have difficulty protecting the environment.

Agencies cannot make business decisions, such as when to open or close a mine, for the operators. However, several states, such as California, have adopted regulations covering temporary closures. California's regulations require the operator, after an operation becomes idle, to submit an interim management plan that must specify how the site will be maintained during the idle period. The interim plan may be effective for not more than 5 years, after which the lead agency must either grant a renewal of the plan or require commencement of reclamation. Unless an interim management plan has been submitted, an operation idle for over one year must commence reclamation. Neither the Forest Service nor BLM have yet adopted regulations dealing with temporary closure. Current low metal prices suggest this can be a serious regulatory issue. Recommendation 5 in Chapter 4 addresses issues related to temporary mine closures.

REGULATORY IMPLEMENTATION ISSUES

When poorly implemented, statutes and regulations are of limited benefit to the environment, the public, and mining operators. The Committee was told of instances in which regulatory agencies, particularly the federal land management agencies, did not implement their statutes and regulations efficiently. Shortcomings included failure to use existing statutory authority to meet environmental protection goals; inadequate information systems to track mining operations and inform the public; unreasonable time delays for NEPA reviews and permits; and inadequate staffing in terms of work loads, skills, and training.

Adequacy of Staff and Other Resources

The Committee found that BLM offices are often unable to meet their field site inspection goals. Some notice-level activities are rarely inspected, and intervals between inspections of operating mines are longer than seems warranted. Oversight of critical mine development projects (e.g., the construction of liners under leach pads or liquor ponds) requires greater staff allocations than some agencies and offices are apparently capable of making. Still, the Committee was impressed by the competence, dedication, and effort expended by many of the agency personnel interviewed.

Staff shortages are likely to be at least partially responsible for the excessive delays experienced in NEPA reviews and issuance of permits that the Committee repeatedly heard described. Actions through the Department of the Interior's Board of Land Appeals are reportedly also unreasonably long. The availability of competent staff was another concern raised during the Committee's study.

Oversight of mining activities is a technologically demanding activity. It often requires hydrogeochemical analysis of proposed tailings ponds, waste ponds, and heap leach pads the hydrologic effects of proposed dewatering schemes; and the long-term impacts of pit waters on surrounding water quality and quantity (see Appendix B). The needed levels of expertise are not always readily available to regulatory agencies. Staff sharing among agencies or between agency offices and use of outside experts is of some, but apparently insufficient, help.

The current use of consultants raises questions about potential conflicts of interest, for in many cases the consultants, although retained by the agency, are paid by the applicant and the applicant therefore retains at least a perceived role in directing the consultant. For instance, most environmental impact statements prepared to support decisions to permit new mining operations or substantial modifications of existing operations are prepared by third parties selected by the agency and the applicant, and paid for by the applicant. Similarly, the agencies have apparently been able to receive technical assistance in spite of budgetary restrictions because the mining companies have offered to pay for the required consultants. The Committee heard of no instance in which this system created a problem, and it does allow the agencies to have access to expertise and resources that would be unavailable to them otherwise. The agencies also have the authority to reject advice and require changes to in draft reports, and have exercised this authority. However, the relationship does create the appearance, if not the fact, of a potential conflict of interest.

The Committee also received evidence that staff training is inadequate, for example, in the interpretation of the agency's regulatory authority and enforcement options. Issues of staff numbers and competence will more seriously affect environmental protection and regulatory efficiency if agencies are required to further reduce the resources devoted to regulating mining activities or if the regulatory burden is increased.

In addition to staffing issues, some administrative and operating procedures of the agencies appear to be out of date and inefficient. Opportunities for more efficient and effective monitoring and enforcement through remote-sensing techniques and air surveillance are worthy of further consideration. Recommendation 12 in Chapter 4 addresses issues related to adequacy of staff and other resources.

Agency Data and Information Systems

The Committee was consistently frustrated by the inability of federal land management agencies to provide timely, accurate information regarding how they manage their lands and the status of mining projects under their jurisdiction. The agencies could provide only approximate information regarding protected lands under their jurisdiction, the area of land currently subject to mining claims, the area covered by land use plans, and other basic land use statistics.

Information about current mining activities was even scarcer. The lack of information appeared to be greatest among highly placed officials who have the greatest need to know. Consequently, those responsible for regulatory management and change, and for keeping the public and Congress adequately informed, appear to be severely limited in their ability to do so.

More specifically, the Committee found it difficult to obtain comprehensive information on agency failures to prevent unnecessary or undue degradation on federal lands. Some compilations of non-compliance were obtained from the BLM, but there was no analysis or interpretation of the data to identify gaps or weaknesses in the regulatory system. The Committee considers this type of information essential for developing any proposals for regulatory change. Federal land management agencies have a responsibility to track the compliance of mines under their jurisdiction and to communicate this information to the public and other stakeholders. Although the Committee got the impression from field-level personnel that such tracking was generally accomplished, the Committee found no systematic, easily available compilation and analysis of the information for public and other stakeholder use.

Regulation of mining on federal lands requires that the federal land management agencies coordinate with a multitude of other regulatory agencies. It was not clear, however, if the federal land management agencies were systematically tracking operator compliance with the commitments made in permits and effectively communicating these findings to the various federal and state regulatory agencies, and to interested stakeholders. Examples of information that could be usefully tracked, compiled, and reported include:

- an up-to-date accounting of the number, commodity, type (e.g., open-pit, underground), status (e.g., active, post-closure), land ownership status, and approximate production of all hardrock mines;
- inspection reports of federal land managers and of other federal, state, and local agencies;
- summaries and analysis of monitoring data and reports that characterize the degree of compliance with standards and criteria at the mine site and highlight incidences of noncompliance;
- appropriate responses taken to return sites to compliance;
- enforcement actions required to assure implementation of required corrective response;
- status of reclamation at the sites;
- status of financial assurances;
- any financial assurance forfeitures required; and
- any outstanding incidents of noncompliance requiring further action.

Reports on the status of NEPA reviews and permit actions, including lapsed time and estimated completion dates, would also be appropriate. Compiled information should be summarized periodically and reported upward in the federal land management agencies. This same information should be readily accessible to the public either as published reports or postings on the Internet. Periodic public reporting of NEPA and permitting status and the compliance status of individual mines, as well as aggregated district, state, and national compliance information, would help the public to determine whether the environment is being adequately protected and whether the regulations are being managed efficiently. This reporting would also focus public and agency attention on those circumstances requiring improvement. Recommendation 11 in Chapter 4 addresses the creation and management of a management information system.

Enforcement Authority

The Committee heard from staff in BLM and Forest Service field offices regarding the adequacy and availability of their enforcement tools for violations of plans of operations (see 43 CFR 3809.3-2 and 36 CFR 228.7[b]). Committee members were struck by the differences in understanding that agency staff members expressed concerning their authority. Staff members in some Forest Service and BLM districts lamented their inability to use enforcement tools effectively, whereas staff members in another district of the same agency expressed confidence that they could use them.

For example, federal land managers can issue a notice of non-compliance to an operator, but must rely on a court injunction to compel an operator to respond. Whereas some districts of the BLM and Forest Service have found the notice relatively useful, others regard it as ineffective. Part of the difference appears to be the ease with which an office can obtain assistance from the U.S. attorney. This depends, in turn, on such factors as location of the federal district court, attorney and agency workloads, agency staff experience, and prosecutor priorities. The agencies have no authority to impose administrative penalties or other sanctions for failures to comply promptly with a notice of noncompliance under current regulations. There is, moreover, some debate about whether the general provisions for regulations under FLPMA or the Forest Service Organic Act authorize the federal agencies to adopt administrative civil penalties in the absence of express statutory provisions for such penalties.

Forest Service staff members advised the Committee that they commonly resort to regulations at 36 CFR part 261 to enforce the terms of operating plans, rather than seeking civil injunctive relief. These regulations contain a general provision that makes violation of a Forest Service-approved operating plan a petty offense punishable by a fine of not more than $500 or six months imprisonment (36 CFR 261.10[l], 261.1b). A $500 fine provides only a limited inducement for operators to correct problems in cases where compliance costs can be several orders of magnitude higher. Access to a U.S. magistrate, assisted by Forest Service law enforcement personnel, apparently is a stronger tool to achieve compliance from operators that are in violation of their plans of operations. Some Forest Service staff members also noted that their use of the 261 authority rather than the mining-specific 228 authority was in part due to poor wording of 228.7(b). That section currently provides that a notice of noncompliance will be issued when an operator is failing to comply with regulations or the plan of operations, and the noncompliance is unnecessarily or unreasonably causing injury, loss, or damage to surface resources. The apparent requirement that actual harm be occurring in addition to a violation has led to concern about the efficacy of the notice of noncompliance in preventing harm to land resources, rather than just correcting damages after the fact.

States retain much of the enforcement responsibility related to mining activities on federal lands. State regulatory programs contain enforcement provisions that provide for issuing notices of non-compliance and assessing penalties when the notices are not resolved. Typically, state agency personnel can issue the notices, and in some cases also the penalties. In other cases, an independent board or panel decides violations and assesses penalties. The state processes for dealing with noncompliance and an operator's failure to respond, which vary widely, may be more timely than the federal processes, which depend on the involvement of a U.S. magistrate.

In addition to these federal and state enforcement authorities, there are enforcement authorities for other federal regulatory programs, such as those of the Army Corps of Engineers and the EPA. These programs may provide a more expedited path for dealing with operator noncompliance than the federal land management agency process. However, the Committee found that federal agency procedures for deciding when to refer apparent violations to state environmental permitting agencies, EPA, or the Army Corps of Engineers are not consistent or readily accessible. Memoranda of understanding (MOUs) address this issue in some states. MOUs do not always address enforcement coordination with federal agencies in those states where EPA (rather than the state) has permitting authority.

While enforcement actions taken by BLM or the Forest Service are generally limited to violations of plans of operations, a plan is likely to include areas that state permitting agencies consider their responsibility. The regulation of water quantity through water rights issued by a state engineer is one such area. Initiation of an enforcement action by BLM or the Forest Service without prior consultation with, and agreement by, a state water resources agency undoubtedly would result in dispute.

In summary, enforcement of mining-related compliance on federal lands is inconsistent and unnecessarily difficult. Recommendations 6 and 15 in Chapter 4 address the enforcement authority of the land management agencies.

SCIENTIFIC ISSUES

Regulatory programs are based on scientific understanding and analyses, and the adequacy of any regulatory program is substantially determined by the adequacy of its scientific underpinnings. The Committee identified several environmental issues whose solutions require more scientific information and understanding than is now available (see Appendix D). Science has contributed greatly to the amelioration of environmental problems associated with mining, and it has the potential to do much more. It serves as the foundation on which to base effective regulations, provides baselines and mileposts for performance expectations, and promises an ever improving ability to forecast and control future environmental impacts. Scientific research and observations have provided the basis for the development of models that predict the concentration of metals and other constituents in leachate and pit lakes over time; for predictions of the impacts of discharge waters on aquatic and terrestrial biota; and for improvements in the effectiveness and environmental protection of exploration and development techniques. A coordinated research effort is needed to properly calibrate and improve the methods and models, and to adequately address such issues as the long-term cumulative effects of mining. Expanded discussions of some of these issues and others related to the environmental impacts of hardrock mining are found in Appendixes A, B, and D.

Water

Among the various environmental concerns related to hardrock mining, those involving the quantity, quality, and distribution of water are dominant. Water quality issues at mine sites may include metals, cyanide, acid drainage, pit lake water quality, placer mining, and mine water discharge to streams, all such conditions being site specific. Water quantity issues include mine water discharge to streams, groundwater withdrawal, and pit lake water quantity. Water quality and quantity issues affect riparian resources and aquatic biota. The potential environmental impacts of mining related to water are discussed in Appendix B. Some of the technological approaches to minimize impacts to water from mining are discussed in Appendix A.

Sidebar 3-1 describes the origin of metal-bearing acid drainage whose prediction and mitigation are critical to planning an environmentally acceptable mining operation. Appendix B includes a discussion of predictive and empirical tests for acid-generating potential. Appendix D addresses some of the uncertainties associated with the accurate prediction of acid mine drainage.

Pit lakes have recently been recognized as major long-term environmental concerns (see Appendixes B and D), and the Committee was able to visit several mines in Nevada that do or will contain pit lakes. The ability to accurately model the future hydrology and chemistry of these lakes and to forecast their roles as social and ecological assets or liabilities is a significant challenge deserving a large and diverse research effort (see Appendix D). The challenge is made even greater by the states that have not classified or designated a status for pit lakes, either individually or collectively. Recommendation 8 in Chapter 4 proposes the funding of a research program for the environmental impacts of mining.

Baselines and Backgrounds

Mine exploration and development occur in remote places where the environment has been relatively undisturbed or in areas that have been disturbed by prior mining or other development. In both cases, data must be collected to establish baseline conditions for tracking the effects of mining on

SIDEBAR 3-1 Where does acid mine drainage come from?

One major environmental concern at many hardrock mine sites is metal-rich acid drainage, often referred to as acid mine drainage. A common feature of most ores is the presence of large quantities of sulfur, usually as pyrite or marcasite (both FeS_2) or pyrrhotite ($Fe_{1-x}S$). These minerals often occur with copper, zinc, lead, silver, arsenic, and other sulfides that constitute the economic part of the mineralization in sulfide ore bodies. Disseminated pyrite, in particular, commonly occurs in areas well beyond the zone of economic mineralization and as a component of waste rock. In such sulfur-rich deposits sulfur is released by the slow natural processes of oxidation, weathering, and erosion. The mountain ranges of the West are populated by areas stained red by the weathering of pyrite to earthy iron oxides, or "rust." The sulfuric acid concurrently produced has been diluted and dispersed, creating local, and generally minor, natural degradation, although some natural acid springs do exist. Nature has created this natural oxidation at a leisurely pace. Mining exposes sulfur-rich material to the atmosphere at a much faster pace, which leads to more rapid oxidation. In the process, mining may create environmental problems with acidic runoff if this sulfuric acid is released to the environment. The acidity of the water and its proximity to metals in the ore may generate waters of low pH that are high in copper, iron, zinc, aluminum, arsenic, selenium, and other elements.

environmental conditions and to help guide reclamation decisions. Baseline measurements are needed on parameters such as rainfall and its variability, infiltration and stream discharge rates, vegetational communities, and the character and dynamics of terrestrial wildlife and aquatic biota over a sufficiently long period to establish meaningful baseline conditions. In previously disturbed areas, such features as the original drainage patterns and soil and water chemistry may be indecipherable. Determining the sufficiency of data is important and difficult for both administration and science, but it is an action that deserves both a practical resolution and continued investigation. The importance of monitoring is discussed in Chapter 1 and Appendix B.

THE NEED FOR EARLY STAKEHOLDER CONSULTATION

The Committee heard from many individuals and organizations that early consultation among all stakeholders is essential for regulatory efficiency. Many individuals who addressed the Committee on this issue emphasized that the NEPA and permitting processes could be expedited if all stakeholders were to participate in the earliest scoping and pre-application meetings. Agreement might not be reached among all of the stakeholders. However, the issues would be better understood by the public and defined to the benefit of the public, the agencies, and the applicant if early consultation occurred under the NEPA and permitting processes. Early consultation should include all stakeholders, including the relevant federal, state, and county agencies, tribes, citizens groups, and the applicant. The Arizona Aquifer Protection Permits Application Guidance Manual provides an example of procedures for pre-application meetings and coordination, in this case between the Arizona Department of Environmental Quality and the applicant.

The Committee heard that some federal agencies had declined to be involved in the scoping and pre-permitting meetings, but subsequently had imposed additional information gathering and other requirements on the applicant. This approach is particularly disruptive and likely has the effect of prolonging both the NEPA and permitting processes. The EPA was repeatedly criticized at the Committee's public meetings, as well as in various reports, for declining to be a cooperating agency in the preparation of mining EISs (EPA, 1997a,b).

The federal land management agencies appear to be attempting to promote early consultation among the agencies most involved in the approval of the proposed mining operations. They testified that they, as a matter of course, encourage many of these agencies to cooperate in the preparation of the EIS. In some cases, however, they have not been sufficiently aggressive in identifying and inviting nongovernment organizations to cooperate in the EIS process. These organizations must have the opportunity to identify their concerns during the scoping process. Otherwise, they may raise their concerns late in the process or in appeals of the final EIS to federal courts, which disrupts the permitting process and makes it difficult to collect the appropriate baseline data.

The land management agencies can invite, encourage, and facilitate the early participation of other federal, state, and local government agencies, but they cannot require it. Recommendation 10 in Chapter 4 addresses the need for early stakeholder involvement in EIS scoping, preparation, and review.

RECLAMATION, CLOSURE, AND POST-CLOSURE MANAGEMENT

The Committee learned of concerns about reclamation, closure, and post-closure management of mining sites and also inspected examples of reclamation at active and inactive mining sites. This section addresses the principal issues related to these topics. Although most of these issues do not lead to specific recommendations, long-term post-closure management of mine sites is addressed in Recommendation 14 in Chapter 4.

Backfilling

The matter of requiring backfilling at open-pit mine sites was addressed in 1979 by a National Research Council report by the Committee on Surface Mining and Reclamation (COSMAR) as follows (NRC, 1979, p. xxviii):

> The [Surface Mining Control and Reclamation] Act requires that [coal-mined] land be restored to approximately its original contours. This provision is generally not technically feasible for non-coal minerals, or has limited value because it is impractical, inappropriate, or economically unsound . . .

> Further, to restore the original contour where massive ore bodies have been mined by the open-pit method could incur costs roughly equal to the original costs of mining. Although technically possible, such backfilling of a large open pit would be of uncertain environmental and social benefit, and it would be economically impractical to mine some deposits under the current cost structures.

The appropriateness of backfilling open-pit mines continues to be a matter of public debate. The Committee has no strong basis to contradict the COSMAR conclusion on backfilling, which was based on an analysis of estimated costs. Although the Committee believes partial or complete backfilling can be environmentally and economically desirable in some circumstances, it was unable to find a basis to establish a general presumption either for or against backfilling in all cases. The NEPA process is appropriate for considering and weighing the costs and benefits of backfilling in a site-specific context.

Federal land management agencies can assist that process with guidance documents that address the following factors:

- the impact of backfilling on the potential availability of mineral resources that may become economically recoverable in the future;

- the site-specific environmental benefits that complete or partial backfilling may provide, such as reducing acid rock drainage, avoiding the development of poor water quality in a pit lake, or protecting surrounding groundwater resources;
- the negative environmental impacts that backfilling may cause, such as delayed reclamation and habitat development; and
- the degradation of groundwater quality if the backfill material is leached or chemically transformed as a result of geochemical conditions in the backfilled pit or underground workings.

The circumstances under which backfilling is most likely to be viable include the following:

- mining areas where multiple ore bodies allow mining and backfilling to proceed without double handling of the backfill material;
- locations where backfilling may eliminate negative environmental impacts, such as acid drainage; and
- sites where backfilling provides an economically viable means of achieving reclamation goals or protecting other specified resource values.

Reclamation Planning for Possible Future Uses

Some presenters to the Committee pointed out that mining operations may create features that are positive contributions to the surrounding area. For example, historic mining features may be of significant economic and cultural value to adjacent communities, either as tourist attractions or as part of the cultural heritage of the communities. Historic mining towns such as Cripple Creek and Leadville, Colorado, are important parts of America's cultural legacy, as well as continued centers of economic activity from tourism. Similarly, modern mines, if appropriately designed and reclaimed, could become important historic or cultural resources in the future. It was also pointed out that portions of a mine, such as waste rock dumps, tailings impoundments, pit lakes and highwalls, and underground mine openings may be attractive to wildlife if appropriately reclaimed. Pit lakes can provide aquatic resources for wildlife in otherwise arid regions of the country. The highwalls of the former open pits offer habitat for raptors. Waste rock dumps and exposed mineral deposits also provide recreational opportunities for hobbyists. And, as the number of operating mines declines, the opportunity for "hands on" training of earth science students increasingly rests with historic mines. Finally, the Committee was told of the responsibility of the United States, as the largest consumer of minerals in the world, to retain opportunities for mining on public lands. Experience has indicated that the most likely locations for future development and extraction of additional mineral resources will probably be in existing mining regions. Based on these arguments, reclamation decisions should carefully consider possible future uses of mine sites. Protection of the environment and consideration of safety are paramount, but those goals may still be met while considering other potential future values and uses for the mine sites.

Post-Closure Issues

One of the most common concerns expressed to the Committee was the post-closure condition and use of past mining sites. It was pointed out that many reclaimed sites will require some degree of monitoring and maintenance by the landowner over the long term. In the case of unpatented mining claims on federal lands, the long-term landowner will be BLM or the Forest Service. Unless other provisions have been made through financial assurance, the costs of continuing care will be paid by the public.

An important part of long-term management will be monitoring, inspection, and low-level maintenance of reclamation features, such as soil covers, vegetation, closed impoundments, waste rock piles, and water diversion structures. In some cases the quality of surface water or groundwater must also be monitored.

It is also important for the federal land management agencies to plan for post-closure uses of the land. Pit lakes, for example, may provide an opportunity for boating and other water sports. Leach and wasterock piles may attract off-road vehicle enthusiasts, and open-pit highwalls may attract weekend rock climbers.

Some of these post-closure uses, however, may violate the integrity of the reclamation features. Extensive off-road vehicle use, for example, could cause significant damage to soil and vegetative covers on leach and waste rock piles. If this occurs, the entry of water and air may exceed the design objectives, possibly leading to unexpected seepage and changes in water quality. Some unintended uses may also conflict with the intended uses of the site. Some recreational activities, for instance, could conflict with intended grazing or wildlife use. Post-closure land use conflicts have the potential for creating significant environmental and safety problems. Dealing with these problems and attempting to manage the unplanned or unforeseen uses may impose additional costs and staffing requirements on the federal land management agencies.

Post-closure problems could also arise from unusual climatic or hydrological events. Climate variations, especially in a desert environment, could lead to flooding or drought, either of which could damage the reclaimed site and result in increased costs to maintain the integrity of the reclamation design and structures. Although the likelihood of extreme events occurring at any one site may be low, the probability that they will occur at some site at some time is high.

Long-term monitoring to detect potential problems and trends must be a part of post-closure plans. The responsibility and costs for monitoring and responding to unexpected uses and events will ultimately remain with the property owner. Ideally, such monitoring consists of nothing more than periodic inspections of the site by the owners or appropriate agency personnel or both. Some sites, however, will require a more extensive and sophisticated post-closure monitoring program.

Monitoring is used for at least the following three purposes:

(1) determine if certain management actions were properly fulfilled;
(2) test predicted outcomes of closure plans; and
(3) identify and characterize trends or changes in site or regional attributes.

Many types of monitoring are important in post-closure issues. The most obvious is monitoring for geotechnical, hydrologic, and geochemical stability. A variety of tools exist for monitoring in each of these areas, and the tools must be selected and used appropriately. Guidance documents and manuals should be developed by the federal land management agencies, and appropriate training should be provided to responsible personnel. Monitoring results should be used by land managers to verify that the outcomes of closure designs are those that were predicted. If monitoring results deviate from the expected outcomes, monitoring should be increased to verify the deviations, and appropriate responses to correct noncompliance should be taken in a timely fashion. Accurate evaluation and interpretation of monitoring data are not trivial tasks, and agencies should be certain that the personnel who are assigned those tasks are well qualified and trained. In some cases, it might be appropriate to employ non-agency professionals to assist in the evaluation and interpretation of the monitoring results (see Appendix B).

Information presented to the Committee suggests that issues of reclamation, closure, and post-closure management have not been adequately addressed by the federal land management agencies. Without changes in planning, the agencies will not be prepared to handle these significant use and environmental challenges, and the public may be faced with major financial liabilities.

REGULATORY EFFICIENCY ISSUES

The charge to the Committee includes provisions to review time requirements, efficiency, duplication, and delay. Specifically, the Committee was asked to address the time requirements applicable to project environmental review and permitting; and to make recommendations and conclusions regarding how permitting requirements and programs can be coordinated to increase efficiency, avoid duplication and delay, and identify the most cost-effective manner for implementation.

Concerns about timing and delays were raised by many who communicated with the Committee. Such delays are an indication of regulatory inefficiency, whether the delay results from inadequate resources or from inefficient use of the resources available.

The Committee did not conduct a management audit of the efficiency of the federal land management agencies, considering that to be beyond the scope of this study. However, comments regarding timeliness and delays are within the Committee's charge. The most serious matter is the excessive time required to complete a NEPA review, followed by the often lengthy time required to issue operating permits.

The laws and regulations applicable to a proposed mining operation and the associated administrative procedures ensure that approval of a new mine is a very deliberate process. In many cases, two years of baseline data are needed before a plan of operations can be submitted. If the data are not available from an existing source, then the applicant may need to wait two years while the data are collected. However, this process has become much slower and more costly than was originally intended or than it needs to be. The process can now take many years. It commonly imposes data collection and analysis requirements on the applicant and the regulatory agency that are poorly coordinated, excessively expensive, and of uneven value in protecting the environment. Mining operators are entitled to a NEPA and permitting process that is as timely and cost effective as possible while still achieving compliance with all statutes and regulations. The federal land management agencies are responsible for making certain this happens by managing the NEPA and permitting process efficiently. The public is entitled to confidence that the foregoing has occurred. Based on presentations to the Committee, improvements in efficiency of the process are needed.

Issues discussed earlier in this chapter affect regulatory efficiency. For example, the early involvement of all key agencies and other cooperating entities in the NEPA process could speed the permitting and make agency resources available for other regulatory work. Freeing staff from the distractions of financially unprotected notice-level activities and small mines and mills without plans of operations, along with providing an adequate management information system, should improve service to clients and make the operations more efficient. Recommendation 16 in Chapter 4 addresses the need for a more timely permitting process that still protects the environment.

PUBLIC INVOLVEMENT ISSUES

The Committee heard evidence of public distrust of the management of mining on federal lands. Local residents near a mine spoke of being excluded from mining decisions that affected the health and quality of life in their community. Others alleged that agencies collaborated with mining operators to advance mine plans regardless of contrary information and views. People spoke of repeatedly being excluded from crucial decisions. Commonly, this was because the decisions were considered to be "technical amendments" to the operating plan or because the proposal was considered to be of such limited significance that it could be addressed through the administrative process, categorical exclusion, or environmental assessment, all of which involve reduced public review and comment.

As addressed earlier, the Committee was frustrated in its attempt to find publicly accessible information regarding mining on federal lands. Such experiences produce distrust in the fairness and propriety of the management and regulatory system. At the same time, the Committee saw evidence of stakeholders developing communication and participation among all interested constituencies. Discussions with federal and state agency employees at state and field office levels generally demonstrated personal and agency commitments to public participation. The committee visited one mining operation where the company took extra steps to keep the public informed by establishing a citizens advisory committee and issuing periodic newsletters about mine activities. These initiatives appear to have raised public involvement and public trust. The NEPA process is designed to promote public participation in decisions regarding federal lands, and it appears to be achieving many of its objectives.

Mining on public lands is controversial and many people hold strongly opposing views. Some people value mining for what it contributes to the national economy and to war efforts. Others are more concerned with pollution and environmental issues than with mineral production. Both sets of values are important and are needed in the debate, but a shift toward more environmental concern is causing difficulties for federal land management agencies as they try to adjust to these changes in public values.

Federal land management agencies historically have given support to such public land uses as mining, logging, and grazing. Now they are giving as much attention to such issues as watershed protection and clean water for public use, wildlife habitat, wilderness, ecosystem preservation, and endangered species. Because these are public lands, they must support the full range of public values. To fulfill this responsibility they must provide information on all aspects of the public lands and encourage the public to use that information in expressing its interests and concerns.

The public is entitled to confidence that public lands are environmentally protected during exploration and mining activities. Federal and state laws, regulations, policies, and procedures go far in assuring this protection, but they are not sufficient. Changes in public values, environmental understanding, mining and reclamation technology, and the federal land management and regulatory agencies demand an extraordinary effort to promote public education, involvement, scrutiny, and trust in the coordination of environmental protection and management of mining on federal lands. The foregoing cannot be achieved solely through public relations programs. Public confidence also cannot be achieved with the currently inadequate management information systems, as discussed in earlier sections. The NEPA process is well suited to help achieve the objectives, but its effectiveness suffers from inadequate participation by many agencies and constituencies. Public confidence in the land management agencies is compromised if the public lacks the ability to track compliance with land use decisions.

The Committee believes that the recommendations in this report, if adopted, will significantly benefit public education and participation, and contribute to a healthy balance between mining and the environment. Recommendations 11 and 13 in the next chapter address issues related to public participation and availability of information about mining operations.

SUMMARY

The analysis of hardrock mining-related issues in this chapter leads to the Committee's conclusions and recommendations presented in Chapter 4, which are based on the elements needed for an effective regulatory program as outlined in the introduction of this chapter. The analysis shows a need for correcting some gaps and inadequacies in the regulations themselves, for improving implementation of the regulations, and for increasing the availability and quality of scientific information. Ignoring these needs for corrections and improvements in the face of rapidly changing technology, economics, and environmental concerns puts the environment, the public, and the mining industry at risk.

4

Conclusions and Recommendations

> *The study shall identify and consider . . . recommendations and conclusions regarding how federal and state environmental, reclamation and permitting requirements and programs can be coordinated to ensure environmental protection, increase efficiency, avoid duplication and delay, and identify the most cost-effective manner for implementation.*

The charge to the Committee has three major components. First, the Committee was asked to identify federal and state statutes and regulations applicable to environmental protection of public lands in connection with mining activities. Second, the Committee was charged with considering the adequacy of statutes and regulations to prevent unnecessary or undue degradation of the public lands. These were discussed in Chapters 2 and 3 respectively. In this chapter, the Committee addresses the third part of its charge and presents its conclusions and recommendations for the coordination of federal and state regulations to ensure environmental protection, increase efficiency, avoid duplication and delay, and identify the most cost-effective manner for implementation.

Each recommendation is followed by supporting text that provides the justification for the recommendation, discusses its implications, and explains the Committee's view of how it can be implemented.

CONCLUSIONS

Existing regulations are generally well coordinated, although some changes are necessary. The overall structure of the federal and state laws and regulations that provide mining-related environmental protection is complicated, but generally effective. The structure reflects regulatory responses to geographical differences in mineral distribution among the states, as well as the diversity of site-specific environmental conditions. It also reflects the unique and overlapping federal and state responsibilities. Conclusions that address overall environmental protection and program efficiency related to technical issues, regulations, or guidance include:

(a) Federal land management agencies' regulatory standards for mining should continue to focus on the clear statement of management goals rather than on defining inflexible, technically prescriptive standards. Simple "one-size-fits-all" solutions are impractical because mining confronts too great an assortment of site-specific technical, environmental, and social conditions. Each proposed mining operation should be examined on its own merits.

(b) If backfilling of mines is to be considered, it should be determined on a case-by-case basis, as was concluded by the Committee on Surface Mining and Reclamation (COSMAR) report (NRC, 1979). Site-specific conditions are too variable for prescriptive regulation.

(c) The Bureau of Land Management (BLM) and the Forest Service need not have identical regulations, but some changes are warranted. The two agencies have broadly similar land management mandates. There are, however, some differences in the kinds of lands they manage, in their specific responsibilities, and in their organization. Whereas some of the Committee's recommendations would make the agencies' approaches to regulating hardrock mining more similar, the Committee is not suggesting that uniformity in all aspects is necessary.

(d) Some small mining and milling operations present environmental risks and potential financial liabilities for the public. These exposures are small by comparison to large operations, but as currently regulated they constitute a disproportionate share of the problems for the land management agencies.

(e) Current regulations do not provide land management agencies with straightforward procedures for modification of plans of operations even with compelling environmental justification.

(f) Federal criteria do not distinguish between temporarily idle mines and abandoned operations. This distinction is important because mines that become temporarily idle in response to cyclical metal prices and other factors need to be stabilized but not reclaimed, whereas mines that are permanently idle need to be reclaimed.

(g) Financial risks to the public and environmental risks to the land exist whenever secure financial assurances are lacking.

(h) Current regulations discourage reclamation of abandoned mine sites by new mine operators. New mineral deposits are commonly found at the sites of earlier mines. Even though the operator of a new mine may volunteer to clean up previous degradation, the long-term liability acquired under current regulations can be significant. As a result, non-taxpayer supported reclamation opportunities are missed and undisturbed lands may be preferentially disturbed for new mining sites.

(i) Post-mining land use and environmental protection are inadequately addressed by both agencies and applicants. The regulations and plans of operation generally specify what actions will be taken to protect water quality and what surface reclamation is to be performed for closure. However, there is inadequate consideration of protection of the reclaimed land from future adverse uses; of very long-term or perpetual site maintenance; or of rare, but inevitable, natural emergencies.

Improvements in the implementation of existing regulations present the greatest opportunity for improving environmental protection and the efficiency of the regulatory process. Federal land management agencies already have at their disposal an array of statutes and regulations that for the most part assure environmentally responsible resource development, but these tools are unevenly and sometimes inexpertly applied. Specific conclusions regarding implementation are as follows.

(a) The National Environmental Policy Act (NEPA) process is the key to establishing an effective balance between mineral development and environmental protection. The effectiveness of NEPA depends on full participation of all stakeholders throughout the NEPA process. Unfortunately this rarely happens in a timely fashion.

(b) The Committee was consistently frustrated by the lack of reliable information on mining on federal lands. The lack of thorough information extends from that needed to characterize the lands available for mineral development to that needed to track mining and compliance with regulations. Without more and better information, it is difficult to manage federal lands properly and assure the public that its interests are protected.

(c) Deficiencies in both staff size and training were observed by the Committee in some offices of land management agencies. Increases in staffing and improved training should result in improved environmental protection and program efficiency.

(d) Forest Service permitting procedures for mineral exploration projects with limited environmental impact commonly take significantly longer than is necessary.

(e) Misunderstandings of the term "unnecessary or undue degradation" (FLPMA, 1976 [43 U.S.C. §§7401 et seq.]) leave some BLM field staff uncertain whether the agency has the authority to protect valuable resources, such as riparian habitats, that may not be specifically protected by other laws.

(f) Federal land management agency representatives are inconsistent in their understanding of their enforcement authority and tools. This results from uncertain interpretations of the statutes and regulations, inadequate staff training, and deficiencies in the tools themselves.

(g) Inefficiencies and time delays in the completion of environmental review under NEPA, issuance of permits, and conduct of other administrative actions unnecessarily consume the resources and time of many stakeholders.

(h) Better information on federal lands is needed to make wise land use decisions. The land use planning process required for BLM and Forest Service lands by the Federal Lands Policy and Management Act and the National Forest Management Act, respectively, provide for identification of land and resources deserving special environmental concern.

Successful environmental protection is based on sound science. Improvements are needed in the development of more accurate predictive models and tools and of more reliable prevention, protection, reclamation, and monitoring strategies at mine sites. The science base is far from complete and environmental protection requires that improvements continue to be devised. Some of the most important environmental concerns at hardrock mining sites are those related to long-term water quality and water quantity, which affect riparian, aquatic biological, groundwater, and surface water resources. A broadly coordinated, national research effort is needed to guide future development and to create improved methods for predicting, measuring, and mitigating environmental impacts related to hardrock mining.

Portions of the public and the mining industry have little confidence in the propriety or fairness of the regulatory and permitting system. Some members of the public perceive that regulators work too closely with the companies and permit operations without sufficient environmental safeguards. Conversely, some mining operators experience delays that they perceive to be caused, in part, by members of the public who seek to forestall mining through the permitting and regulatory processes. Lack of confidence in the regulatory and permitting system can lead to delays and higher costs for industry, regulatory agencies, and the public and can also limit opportunities for improving environmental protection.

Conditions are changing for regulations and mining. Technology, social values, the economy, and scientific understanding change continually. Thus, environmental regulations applicable to mining will be most effective if they can use these changes to improve environmental protection. Similarly, the mining industry should benefit through lower operating cost and greater environmental protection. Therefore, a regulatory system that is adaptive to change will serve the public, the environment, and industry best.

RECOMMENDATIONS

Recommendation 1: Financial assurance should be required for reclamation of disturbances to the environment caused by all mining activities beyond those classified as casual use, even if the area disturbed is less than 5 acres.

Justification

The Committee observed unreclaimed exploration and mining sites that currently fall under BLM's category of notice-level activities, where the land disturbance amounts to less than 5 acres. Some of these sites are examples of unnecessary or undue degradation. The Committee believes that the inability to require a bond or other financial assurance for reclamation of these sites represents a gap in existing regulations. This recommendation proposes a major change in the way BLM regulates notice-level activities, which currently do not require financial assurances.

According to data reported by BLM, there are 181 currently outstanding notices of noncompliance for 14,989 currently active notice-level activities (BLM, 1999a, Tables 3-2 and 3-6); that is, less than 2% of the currently active notice-level operations have outstanding notices of noncompliance. Approximately 72% of the outstanding notices of noncompliance for notice-level activities are for "failure to reclaim" (BLM, 1999a, Table 3-6). Although the percentage of notice-level activities cited for noncompliance is small, and even fewer sites are left without some form of reclamation, BLM staff spend significant time locating operators of these sites and achieving compliance with reclamation requirements.

Disturbance of surface resources beyond those defined as casual use (see Sidebar 1-3) are considered significant. If bonds or other acceptable types of financial assurances were posted for reclamation of all such disturbances, regulatory agencies could use these funds to reclaim sites that have been abandoned or for which the operator is unwilling to reclaim the land. The requirement of financial assurance would protect taxpayers, who might otherwise be shouldered with the responsibility of reclamation. Furthermore, the bond itself provides an incentive for operators to reclaim the land in a timely manner.

Discussion

The objective of this recommendation is to guarantee financial assurance for all significant disturbances, while continuing to allow activities with negligible impact to be conducted without permits. In this context significant disturbance for the purpose of requiring financial assurances generally occurs when motorized equipment is taken off existing roads onto a site. Examples include the use of bulldozers (for building new roads and drill pads) and backhoes (for digging exploration trenches).

The Forest Service currently uses a standard of "significant disturbance" in determining the threshold beyond which a plan of operations is needed (26 CFR 228.4). On BLM lands, neither a notice nor a plan of operations is required if the activity is classified as "casual use," which is an activity ordinarily resulting in only negligible disturbance of the land and its resources (see Sidebar 1-3). BLM generally considers as casual use those activities that do not involve mechanized earth-moving equipment or explosives or do not involve motorized vehicles in areas designated as closed to off-road vehicles (43 CFR 3809.0-5).

Implementation

Standard bond amounts for certain types of activities on specific kinds of terrain should be established by the regulatory agencies. It should be recognized that certain types of activities are less costly to reclaim than others. A set of activity- and terrain-dependent standard bond amounts (by state, BLM district, or forest) should be established for typical activities, especially those of prospectors, small exploration companies, and small miners, so that adequate bonds are posted for activities under 5 acres and so that the permitting process is expedited. Standard bond amounts (a certain number of dollars per acre of land disturbed for a particular type of activity) should be used in lieu of detailed calculations of bond amounts based on the engineering design of a mine or mill. In addition, the Committee encourages the use of bond pools to lessen the financial burden on small miners.

Recommendation 2: Plans of operations should be required for mining and milling operations, other than those classified as casual use or exploration activities, even if the area disturbed is less than 5 acres.

Justification

The Committee observed examples of unnecessary or undue degradation on a few notice-level mining operations, which are not now required by BLM to have plans of operations. The Committee considers this to be a gap in the current regulations. This recommendation proposes a major change in the way notice-level operations are treated.

Mine development, extraction, and mineral processing activities generally disturb the land more significantly than exploration activities (see Appendix A, which describes the overall mining process, Appendix B, which discusses impacts, and Recommendation 3, which addresses notice-level exploration activities). As such, mine development, extraction, and mineral processing require considerable engineering design and construction activities, whereas, apart from the design of roads to minimize erosion and impact on sensitive areas, exploration requires little, if any, engineering and construction. A plan of operations should be subjected to regulatory approval to assess and help minimize environmental disturbance from mine development, extraction, and mineral processing activities. (Recommendation 3 addresses expedited procedures for exploration.)

Discussion

This recommendation seeks to ensure that potential environmental impacts of mine development, extraction, and mineral processing operations are adequately considered and that the land will be reclaimed after mines and mills close.

The Committee favors retaining the BLM distinction for casual use operations and the Forest Service's similar recognition of activities with no significant impact.

The Committee discussed two examples in an effort to determine which kinds of mining-related activities should have approved operating plans. In the first example, small suction dredges used to recover placer gold from sediments in streams generally are allowed under various state laws to be in the streams only during certain times of the year, preventing disturbance of fish at critical stages in their life cycles. The Committee believes that BLM and the Forest Service are appropriately regulating these small suction dredging operations under current regulations as casual use or as causing no significant impact, respectively.

In the second example, the Committee debated whether bulk sampling for metallurgical testing, an activity that could be considered advanced exploration rather than mining, should require a plan of operations or be considered a notice-level activity. Bulk sampling commonly involves excavation, from a shallow open pit or small underground openings, of 10 to 1,000 tons or more of presumed ore. The rock is tested, typically off site at an operating mill or at a metallurgical laboratory, to determine whether the contained metals can be extracted efficiently and profitably. Because an exploration project must advance to a considerable degree before bulk sampling is done and because bulk sampling can require the excavation of considerable amounts of overburden and waste rock, the Committee believes a plan of operation should generally be required for activities involving bulk sampling.

Thus, the Committee believes that, in association with a quick review of the operating plans, these same types of operations would benefit from a NEPA review that uses existing and readily available information. Some of this information may be available through the documentation developed by the land management agency for the land use planning process. The NEPA process will commonly require an environmental assessment rather than an environmental impact statement; in either case, the land management agency is responsible for timely completion of the appropriate NEPA documents. The requirement for a plan of operation for all mine development, extraction, and mineral processing operations should not be viewed as an opportunity to slow the process through extended review, but rather as an opportunity to develop the information needed for improved operations and for better monitoring and enforcement.

Implementation

The Committee recognizes the valuable role that small miners play in the development of the nation's mineral wealth. Because of scale, small mines generally have less potential environmental impact than major mines. The Committee therefore believes that, with adequate bonding for reclamation, small miners should receive expedited schedules and services in permitting. For example, regulators should provide small miners with examples of generic permit applications that clearly explain the miner's responsibilities. The Committee believes that certain types of mineral extraction processes should be treated differently than others in terms of the speed with which plans of operations are reviewed and approved. Some could have rapid, check-off-the-boxes approaches to permit applications. For example, mining for mineral specimens for collectors could be reviewed and permitted rapidly if no potentially harmful chemicals are used on site for beneficiation and if less than 5 acres of land is disturbed. Most specimen mining occurs in established mining districts, where adequate surface access exists. In addition, the land management agencies could assist small miners and mill operators by assigning more personnel to help complete permit applications, educate the operators in the permitting process, and develop standardized, easy-to-understand forms.

Recommendation 3: Forest Service regulations should allow exploration disturbing less than 5 acres to be approved or denied expeditiously, similar to notice-level exploration activities on BLM lands.

Justification

The ability to obtain access to land in a timely manner is vital to exploration (see Appendix A). The Committee found that the time it takes industry to obtain permission to explore on federal lands varies considerably by agency, even when disturbance to the land is likely to be minimal. This recommendation calls for a major change in the way the Forest Service approves exploration activities for disturbances of less than 5 acres. Along with Recommendation 1, which would require financial assurances for such exploration activity on BLM lands, this recommendation would make the requirements for exploration on Forest Service and BLM lands similar.

Under the current system for notice-level exploration activities affecting 5 acres of land or less, BLM has 15 days to respond and notify the operator if extraordinary measures are needed for the planned activities. In contrast, Forest Service officials reported that essentially identical exploration activities on Forest Service lands often require eight months lead time and sometimes as long as two years to obtain approval, although some approvals for exploration are obtained more quickly. Before plans of operations for exploration are approved, Forest Service officials generally inspect the sites, carefully review the plans, and prepare environmental assessments. The length of time can vary depending on whether the operator has the financial resources to provide for a third party to prepare an environmental assessment or must wait for the Forest Service to prepare the document.

Exploration typically progresses from one place to another depending on the results of the previous tests. For example, the results of one drill hole will often dictate the optimal location of the next drill hole. Anticipating the location of all drill holes on a particular piece of land is difficult. Therefore, exploration companies, when planning their exploration programs, tend to add to the estimated number of drill hole locations that are necessary.

Exploration companies informed the Committee that companies typically attempt to limit disturbance to less than 5 acres at a time on BLM lands because of the quick review from BLM for notice-level disturbances. Because such a provision does not exist in Forest Service regulations, there is less incentive to limit disturbance on these lands. In fact, because the review time is so lengthy, companies are likely to submit proposals for far larger exploration programs on Forest Service land than on BLM land in order to avoid further delays that would be caused in obtaining additional approvals. The only incentive for limiting disturbance on Forest Service lands appears to be the cost of reclamation.

Discussion

The objective of this recommendation is to allow exploration activities to be conducted quickly when minimal degradation is likely to occur. The Committee believes that, with reclamation bonds or other financial assurances in hand for land disturbance (see Recommendation 1), exploration should be able to proceed expeditiously. That is, the current BLM 3809 regulation with a 15-day response time for notice-level exploration activities should be maintained, and a similar procedure should be adopted by the Forest Service.

Increased use of geographic information systems to check locations for such features as riparian areas, steep terrain, cultural resources, and known habitats and locations of threatened or endangered species should facilitate the necessary evaluations by land managers. Quick response time improves exploration efficiency because it gives operators the necessary flexibility to modify plans based on initial results, and because it provides an incentive for operators to design exploration activities that minimize the disturbed acreage.

Implementation

Although the Committee recognizes that Forest Service lands often are in steeper terrain with more springs and creeks than BLM lands, the Committee believes that, with financial assurances in place, exploration disturbing less than 5 acres should be able to proceed on Forest Service lands as rapidly as on BLM lands. That is, the Forest Service should modify its 228 regulations to allow for rapid approval of exploration activities disturbing less than 5 acres.

The Committee does not intend that the requirement of bonding for exploration activities (Recommendation 1) result in a federal action that would automatically trigger an environmental assessment or environmental impact statement. BLM does not currently require companies to supply an environmental assessment for notice-level activities. The Committee believes that not requiring environmental assessments for exploration with less than 5 acres of disturbance is appropriate for both BLM and Forest Service lands.

The Committee debated whether the 5-acre threshold for BLM notice-level activities, above which full plans of operations with environmental assessments or environmental impact statements are required, is appropriate. Five acres appears to be a reasonable threshold, if reclamation bonds or other financial assurances are required, because environmental impacts from exploration at this level are likely to be minimal and because the cost (generally $5,000 to $25,000 for a full 5 acres of expected disturbance) is likely to be sufficiently high to provide an incentive for operators to reclaim the land quickly.

Recommendation 4: BLM and the Forest Service should revise their regulations to provide more effective criteria for modifications to plans of operations, where necessary, to protect the federal lands.

Justification

Concern was expressed to the Committee about the ability of BLM and the Forest Service to require modifications of plans of operations in light of new circumstances or information, such as unexpected acid drainage, problems with water balance, adequacy of approved containment structures, or discovery of impacts on wells and springs. The ability to require operators to make necessary modifications is essential to the agencies' ability to prevent unnecessary or undue degradation of federal lands.

A BLM state director or Forest Service supervisor now may require modifications only after making a set of findings that include determinations on the foreseeability of the issue when the original plan was approved, and if the agency took all reasonable measures to forestall the issue at that time. However, arguments over what should have been "foreseen" or whether a BLM or Forest Service officer took "all reasonable measures" in approving the original plan makes the modification process dependent on looking backward. Instead, the process should focus on what may be needed in the future to correct problems that have resulted in harm or threatened harm. Protection of the lands should not depend entirely on an after-the-fact assessment of how good a job was done at the time the original plan was approved. It is more important to address what can be done now to avoid unnecessary or undue impacts. Modification procedures should look forward, rather than backward, and reflect advances in predictive capacity, technical capacity, and mining technology.

Discussion

Disputes have arisen over the ability of the BLM and Forest Service to require modifications under particular circumstances, for example, Island Mountain Protectors, IBLA 97-76R et al. (Nov. 20, 1998) (dispute over whether revised closure plan is or is not a "proposed modification" of the plan of operations); and Creole Corp., 146 IBLA 107 (1998) (dispute over authority to require modification where detailed factual findings were not made on adequacy of 17-year old plan of operations where active operations had never commenced).

In part, this look backward is intended to prevent unfairly reopening issues that were, or should have been, anticipated at the outset. In practice, these requirements can complicate attempts to address current effects that became apparent only after some period of operation. See, for example, Red Thunder, Inc., 129 IBLA 219, 101 I.D. 52 (1994) (recounting dispute over legal status of letter "revisions" to plan of operations when acid drainage had been considered in the original plan of operations, but mining of sulfide ores had not been anticipated).

Various offices of BLM and the Forest Service have resorted to different strategies to preserve flexibility to respond to environmental harm or threatened harm despite the limitations of the current regulations. For example, some BLM offices routinely include reopener provisions in approved plans of operations, explicitly providing for submission of modifications where monitoring or other operating experience shows results different from those that were anticipated. The Forest Service often issues plans of operations that are valid only for a defined period—thus requiring the operator to submit a new plan of operations periodically, giving the Forest Service an opportunity to require modifications in the new plan. Some larger operators have agreed to include periodic review and update requirements in their plans to give the land management agency some ability to make necessary changes. While these approaches do not entirely conform to current regulations, they have helped protect the environment. Staff comments and documents reviewed by the Committee suggest that the regulations should be modified to improve criteria for modifications, require periodic reviews, and/or specify expiration dates for approved plans of operations to assure the opportunity to adjust practices where needed.

Implementation

Provisions for periodic review of plans of operations, and the ability to require modifications, are important to deal with adverse effects on public lands. BLM and Forest Service regulations should not require the agencies to make retrospective findings on "foreseeability" or whether "all reasonable measures" were applied in approving the existing plan. Modifications should be based on the results of monitoring or other data that demonstrate the occurrence or likely occurrence of unnecessary or undue degradation if the plan is not modified.

One of the implications of requiring modifications is the need for NEPA review. Where such modifications are needed to prevent unnecessary or undue degradation, such review should be expeditious and tied to the NEPA document approving the initial plan of operations. In addition, revised agency procedures should contain safeguards to assure that modifications are imposed only after serious consideration and following a procedure that protects the interests of the mining company in continuing to conduct operations, consistent with the avoidance of unnecessary or undue degradation.

The Committee did not determine if plans of operations should be reviewed or reopened at predetermined intervals. The evolutionary nature of mining at individual sites—particularly at mines using newer technologies and dealing with disseminated mineral deposits—requires changes in the limitations on plan modifications in the original BLM and Forest Service regulations. Updating of financial assurance instruments should also take place as conditions change that might affect the levels of bonding or other forms of financial assurance. Practices now vary among the states and federal agencies.

Recommendation 5: BLM and the Forest Service should adopt consistent regulations that a) define the conditions under which mines will be considered to be temporarily closed; b) require that interim management plans be submitted for such periods; and c) define the conditions under which temporary closure becomes permanent and all reclamation and closure requirements must be completed.

Justification

Temporary closures as a result of low mineral prices may cause environmental problems if appropriate management measures are not undertaken. Federal land management agencies should have regulations to address such events. The land management agencies need to have the authority to require an operator to properly close a mine rather than allow it to remain in limbo if poor market conditions persist. The situation today is quite different from that over the last two decades when gold and other mineral prices were high and relatively stable.

Discussion

Under some state regulations, temporary closures (defined as substantial reductions in output) exceeding a specific period of time trigger a requirement for permanent closure. Some companies have used "temporary" status to delay or avoid taking appropriate final actions to clean up and otherwise close their facilities. Agencies should have a clearly stated process and criteria to be used in responding to temporary closures. Several states already have adopted regulations addressing this issue for mining operations.

Implementation

The Forest Service and BLM should develop a common approach to addressing temporary closures and should delineate when temporary closures become permanent. This should be implemented through a rule-making process. They should also enter into understandings, and if necessary formal agreements, with the states that have already adopted such regulations concerning how the agencies will coordinate their efforts and which agency will take the lead in addressing temporary closures.

Recommendation 6: Federal land managers in BLM and the Forest Service should have both (1) authority to issue administrative penalties for violations of their regulatory requirements, subject to appropriate due process, and (2) clear procedures for referring activities to other federal and state agencies for enforcement.

Justification

Field-level BLM and Forest Service personnel told the Committee that they have experienced difficulty, in some cases, in enforcing compliance with regulations and the requirements of notices and plans of operations. The Committee is not able to assess the relative levels of on-the-ground compliance and noncompliance resulting from the limitations of the existing enforcement procedures. However, the Committee is persuaded that more consistent and accessible procedures for deciding when to refer apparent violations to other agencies and the ability to issue reasonable administrative penalties, subject to appropriate due process, would improve the efficiency of agency operations and enhance the protection of the environment.

Discussion

Currently, federal land managers must seek a court injunction to compel an operator to respond to a "notice of noncompliance," which directs the operator to return to compliance (43 CFR 3809.3-2; 36 CFR 228.7[b]). While some districts of BLM and the Forest Service have found such notices sufficient, others regard them as ineffective. The difference appears to be the ease or difficulty with which these offices can obtain the assistance of a U.S. attorney to seek injunctive relief.

Forest Service staff members also advised the Committee that they resort to regulations containing a general provision that makes violation of any Forest Service "approved operating plan" a petty offense punishable by a fine of not more than $500 or six months imprisonment (36 CFR 261.10[l], 261.1b). Access to a U.S. magistrate, assisted by Forest Service law enforcement personnel, apparently gets the early attention of operators that are in violation of their plans of operations.

Some Forest Service staff members also reported the use of the 261 authority, rather than the mining-specific 228 authority, because of poor wording of the "notice of noncompliance" section of 228.7(b). This section currently provides that a notice of noncompliance will be issued when an operator is failing to comply with regulations or the plan of operations, and the noncompliance "is unnecessarily or unreasonably causing injury, loss, or damage to surface resources." The apparent requirement that actual harm be occurring in addition to a violation leads to concern about the efficacy of the notice of noncompliance in preventing harm to forest land resources.

Currently, the consequence for failing to comply with a notice of noncompliance from BLM or the Forest Service is the possibility that an operator may be ordered to comply by a federal court at some later date, with no sanction or disincentive for not responding at the outset. In the case of the Forest Service, petty criminal prosecution is also an option. The Committee finds that an administrative penalty authority should be added to the array of enforcement tools in order to make the notice of noncompliance a credible and expeditious means to secure compliance.

In addition to their own enforcement authorities, federal land management agencies need to acknowledge and to rely on the enforcement authorities of other federal, state, and local agencies as much as possible. The provision of this additional enforcement authority for federal land management agencies makes it imperative that formal understandings be developed between those agencies and the state and federal permitting authorities to prevent duplication of effort and to promote efficiency.

Implementation

BLM and the Forest Service should establish more consistent and accessible procedures for deciding when to refer apparent violations to other agencies that have direct interests in, and responsibilities for, operator activities that adversely affect the environment. Using memoranda of understanding (MOUs) or other means, the procedures should allow for the expeditious involvement of other agencies, except in cases where the land management agency needs to act immediately to protect public lands and resources, or in cases where the other agency is unable or unwilling to act with appropriate speed. In these cases clear procedures and standards are needed to guide the use of existing enforcement authority when violations have occurred or are imminent.

In addition, BLM and the Forest Service should seek authority from Congress, if statutory authorization is necessary, or otherwise revise and expand the existing enforcement provisions in the 3809 and 228 regulations to include administrative penalty authority for violations of their regulations and operating plans or notice-level requirements. Such penalty authority should be subject to appropriate due process, including the opportunity for the operator to rectify the circumstance of noncompliance, to undertake an administrative appeal, and to seek recourse to the courts.

Recommendation 7: Existing environmental laws and regulations should be modified to allow and promote the cleanup of abandoned mine sites in or adjacent to new mine areas without causing mine operators to incur additional environmental liabilities.

Justification

Representatives of several mining companies indicated to the Committee that they would consider partial or full voluntary cleanup of some abandoned sites in or adjacent to their ongoing operations. They were, however, concerned about liabilities they might incur in doing work on abandoned sites. Under the Clean Water Act, CERCLA, and potentially under RCRA, mine operators doing any work at an abandoned site could incur liabilities.

In most cases, some cleanup would improve environmental quality and is better than no cleanup. In addition to federal impediments to voluntary clean up, some states have environmental laws that may pose additional barriers, laws that may be more stringent than their federal counterparts. The Committee did not address the general problem of abandoned hardrock mines on federal lands, but it believes barriers to reclaiming abandoned mines sites in connection with new operations should be removed.

Discussion

The goal of environmental regulations is to protect and improve the environment. There are opportunities for correcting environmental problems caused by abandoned mines, in some cases where new mining is being done in old mining districts. Concern over legal liabilities or the ability to meet regulatory standards leads mine operators to design around older mined areas and pre-existing discharges. In many cases, however, reclamation of previously mined areas would be a reasonable approach for combining construction of the new mine with improvement of environmental problems caused by earlier mining. For example, existing pits might be appropriate places for waste rock disposal; construction of tailings facilities might present opportunities to stabilize or reclaim previous disposal sites; or replacement wetlands sites might be located to provide some treatment for existing poor-quality discharges. Incentives might be needed to assure that appropriate opportunities for reclamation and improvement of environmental impacts are not missed. Similarly, in previously mined areas adjacent to new mining operations, but not necessary for operation of the new mine, there may be opportunities for voluntary activities that could benefit the public. The added cost of full or partial cleanup of these sites could be considerably less than later cleanup under CERCLA rules. For example, use of equipment that is already on site would be less costly for the company or the government than the stand-alone costs if the cleanup were to be accomplished by bringing in the equipment at another time. By working with stakeholders and agencies responsible for overseeing the Clean Water Act, CERCLA, and other potentially inhibiting regulations, federal land management agencies could affect some degree of environmental improvement at old mining sites.

The Committee does not propose these changes to allow cleanup of abandoned mines without any liability for performance of the work. Regulatory agencies must be able to review and approve a work plan, to oversee the work, and to use appropriate mechanisms of control if the actions taken are not improving the environment. The objective of changes in laws and regulations would be to recognize that environmental improvement is worth pursuing at abandoned sites and to limit the liability incurred by the cleanup entity. The Committee also recognizes that if a responsible party can be identified for the site, the site may not be considered truly abandoned and other cleanup approaches can be pursued.

Some states (e.g., Colorado and Montana) have voluntary cleanup regulations for all types of industrial sites. State regulators and the entities that have used those regulations for abandoned mine-site cleanup indicate that these regulations appear to be working well in promoting cleanup of abandoned sites or sites grandfathered from reclamation requirements. However, their application is limited to those sites where the Clean Water Act or CERCLA liability is not a concern. In addition, the ability of regulators to regulate site cleanup varies with each program.

Implementation

To promote voluntary cleanup programs at abandoned mine sites, Congress needs to approve changes to the Clean Water Act and the CERCLA legislation to minimize company liabilities. The federal land management agencies, in cooperation with the EPA, and other state or federal agencies with potential regulatory authority, should propose to Congress changes in regulatory authority that would promote voluntary cleanup. Input should be solicited from other stakeholders, such as mining companies, tribes, and environmental organizations.

Recommendation 8: Congress should fund an aggressive and coordinated research program related to environmental impacts of hardrock mining.

Justification

Regulatory agencies and the mining industry are not adequately addressing research needs related to the environmental aspects of hardrock mining and reclamation, including the uncertainty associated with predictive modeling. The Committee recommends that Congress provide funds to support this important research. The mining industry, environmental groups, and other stakeholders should participate in defining needed research. Some of these needs are identified in Appendix D, including those related to water quality and quantity and to biological and ecological impacts. Research that looks at cumulative impacts, either from multiple mines or multiple uses should be supported to sort out the significance of the impacts of each use. That applies, for example, to cumulative impacts of groundwater withdrawal on regional aquifers and regional water balances.

The potential for long-lasting environmental impacts from mining is high, but research on these issues is inadequate. Ideally, prioritization of and participation in funding for research on these issues should come from a combination of governmental, environmental, and industrial entities.

Discussion

Unlike many other kinds of industrial point-source emissions, releases from mining operations may take years to develop (e.g., acid drainage) and may continue for many years after a mining operation has closed. So far, the success of modeling long-term water quality and quantity impacts has been fragmented, and the concordance of predicted and actual outcomes has not been adequately reviewed. Sampling, testing, and monitoring can provide data to help test, refine, and calibrate long-term predictive models.

The biological aspects of mining are even more neglected and deserve research support. Success in improving environmental controls has shifted some of the most severe remaining impacts of hardrock mining from humans to aquatic and terrestrial biota. Species that may be affected include some threatened or endangered species. Most research on pollution prevention has been focused on manufacturing wastes, but the potential for reducing process waste in mining may exceed that in other industries because of the large scale of mining. Advanced mining technologies, such as bioleaching and *in situ* solution mining, have the potential to reduce pollution and minimize surface disturbance, but the long-term environmental consequences of these methods have not been adequately investigated.

Implementation

Research on the long-term environmental effects of hardrock mining should involve federal agencies (USGS, BLM, USFWS, USFS, EPA, NSF, DOE/National Laboratories), state mining and environmental regulatory agencies, state geologic surveys, universities, environmental groups, and the hardrock mining industry. The capabilities and cooperation of all these groups are needed. Lacking at the present time are adequate funding, a broad commitment to a coordinated research effort, and a focus for coordinating research and disseminating the results. The land management agencies should participate, along with federal and state regulatory agencies, by identifying their priorities for research related to their land management responsibilities.

No single agency or organization can take on full responsibility for the variety of research needs. In addition to the range of government research organizations, each with topical and disciplinary research strengths, some large mining companies have research facilities that could contribute to a focused program. A research program on environmental effects also should have a strong competitive grants program to stimulate relevant academic research. In addition, the EPA and the U.S. Fish and Wildlife Service have research programs that could contribute to a better understanding of the environmental impacts of mining. A listing and discussion of significant research topics is given in Appendix D. The environmental issues are serious enough to warrant immediate congressional funding of research in this area.

Recommendation 9: BLM and the Forest Service should continue to base their permitting decisions on the site-specific evaluation process provided by NEPA. The two land management agencies should continue to use comprehensive performance-based standards rather than using rigid, technically prescriptive standards. The agencies should regularly update technical and policy guidance documents to clarify how statutes and regulations should be interpreted and enforced.

Justification

The intent of this recommendation is to retain the advantages of the current decision-making approach, which bases environmental protection and reclamation requirements on site-specific evaluations conducted pursuant to NEPA, while encouraging the agencies to provide clear public information about their interpretations of and intentions for implementing their regulatory requirements. Several individuals recommended to the Committee that the land management agencies develop regulations incorporating specific numeric standards to control many aspects of the operation and reclamation of hardrock mines. Such an approach would have the advantage of providing explicit quantitative requirements that would be clear to all stakeholders. It would have the disadvantage of reducing the ability of the operating plan and reclamation requirements to be sensitive to site-specific conditions and considerations.

Discussion

The Committee believes that the NEPA process and its various state equivalents provide the most useful and efficient framework for evaluating proposed mining activities for three reasons.

First, the NEPA process provides the most comprehensive and integrated framework for undertaking these evaluations. The NEPA environmental assessment process includes the full range of environmental concerns, whether or not they are specifically addressed by some other regulatory program, as well as cultural and other concerns. It allows for clear identification of tradeoffs between different and sometimes competing values, and promotes a better understanding by all stakeholders of the implications of the many decisions involved in the preparation and approval of a mine's operating plan. It also provides a framework for coordinating the diverse information requirements, concerns, and permitting decisions of the many regulatory agencies and other stakeholders. No other regulatory program provides such a comprehensive, integrated mechanism for decision making.

Second, the NEPA process ensures that the decisions are based on careful analyses of site-specific conditions. For example, ore deposits occur in every conceivable type of geographic and geologic location from arid deserts to tropical rain forests to high mountains. The values and sensitivities associated with diverse types of environmental and cultural resources can vary at least as widely. An operating plan for mining activities must adapt and respond to these specific conditions and sensitivities. The NEPA process allows this responsiveness; regulatory programs relying on inflexible, technically prescriptive standards often do not.

Setting objectives for pit lake water quality is a good example of the advantages of the site-specific NEPA process. Requirements for the long-term quality of pit lake waters vary widely among the states and even in the same state. Water quality standards applied to pit lakes vary from those appropriate to industrial ponds to those required to protect waters of the United States. Pit lake settings are highly variable, and include circumstances where pit waters discharge to downstream watersheds, where pit lake waters mix with surrounding groundwater, and/or where pit lakes are closed basins subject to high evaporation. The Committee concluded that pit lake water quality should be subject to regulation and not simply left to chance. However, the Committee had difficulty identifying a universal approach suitable for the classification of all pit lakes.

In the Committee's view, the site-specific, long-term consequences of pit lakes should be fully considered in the NEPA process. Project approvals should clearly establish acceptable post-closure water quality conditions appropriate to the long-term use of the site and those that provide adequate protection for affected ground and surface waters, as well as wildlife and water fowl.

Third, mining technology for a site can vary substantially, depending on the type of ore, the nature and extent of the ore deposit, and many other site-specific conditions. Mining technologies also have changed, and will continue to change. The NEPA process allows the agencies to be responsive to such technological differences. Less flexible regulatory approaches do not allow this flexibility and, as a result, can cause technologies to be "frozen," often with adverse impacts for both the mining operators and the environment.

For all these reasons, the Committee believes that the agencies should continue to rely to the maximum extent possible on the flexible, comprehensive NEPA evaluation process for making permitting decisions. However, the Committee also recognizes that the NEPA process is not perfect. The process is complex and time consuming and can be implemented inefficiently (see discussions under Recommendations 10 and 16). In addition, although the process recognizes environmental impacts of alternative actions, its objective is to minimize (not necessarily prevent) impacts through mitigation. Therefore some unrestorable impacts may occur even with agreed mitigation measures.

The Committee also believes it is in everyone's best interest if the agencies clearly set forth their policies, interpretations, intentions, and environmental goals in guidance documents and policy papers and make them available to all stakeholders. Discussions with agency personnel demonstrated that many differences in interpretation and perception about existing authorities and policies currently exist in an agency. Such confusion in the agencies can only confuse other stakeholders about the agency's policies. By clearly setting forth these interpretations and policies the agencies can reduce confusion and increase the efficiency of the decision-making process for everyone. By using guidance documents and policy papers rather than regulations for this purpose, the agencies can retain the flexibility that the Committee believes is so important.

Implementation

Continued reliance on the NEPA process requires no modifications to current agency procedures except as identified under other recommendations. The agencies already have the authority to issue guidance documents and policy papers. The challenge is to implement a process for developing, issuing, and updating such documents that allows them to be informed by stakeholder comments (including public comments, where appropriate), while avoiding the inflexibility and costs of the traditional regulatory process. In order to be useful and effective, the advisories also must be easy for mining claimants to obtain. Examples of such guidence documents created by the British Columbia Ministry of Energy and Mines include Price and Errington (1998) and Price (1997).

Recommendation 10: From the earliest stages of the NEPA process, all agencies with jurisdiction over mining operations or affected resources should be required to cooperate effectively in the scoping, preparation, and review of environmental impact assessments for new mines. Tribes and nongovernmental organizations should be encouraged to participate and should participate from the earliest stages.

Justification

The lack of early, consistent cooperation and participation by all the federal, state, and local agencies involved in the NEPA process results in excessive costs, delays, and inefficiencies in the permitting of mining on federal lands. Delayed and inconsistent involvement by nongovernmental organizations exacerbates these problems. In some cases, the agency representatives who do participate in the NEPA process are not those who will be responsible for issuing the ultimate permits, resulting in the raising of new information collection and analytical requirements during the final permitting phases. The objective of this recommendation is to promote efficient, fair, and timely consideration of mining proposals by streamlining the NEPA compliance process and eliminating permitting delay.

Discussion

The Committee received testimony about the frequency of this problem and the substantial costs and delays that it can cause. Examples of problems include:

- failure to achieve early consensus on the appropriate scope and methodologies for collecting baseline environmental data (this failure may result in the need to repeat studies);
- introduction of new issues not included in the initial scoping by parties, including agencies that failed to participate in early consultation;
- late requirements to analyze additional alternatives not included in the original scope;
- intra-agency disconnect between personnel responsible for NEPA compliance and those responsible for permit decisions, resulting in late requests for additional information that should have been included in the original NEPA document; and

- multiple draft documents and revisions requiring additional public review.

The EPA was frequently singled out as an agency that often creates such problems because of its unwillingness to participate early in the NEPA process.[1] Other nongovernmental and tribal organizations were also cited for failing to participate early in the process, partly because the land management agency did not invite or encourage these organizations to participate.

On the other hand, several examples of effective collaboration among agencies were also presented to the Committee. Cooperation appeared to work particularly well when the state also had an environmental impact assessment requirement. The earlier the joint efforts begin the more effective and efficient the decision-making process is for everyone concerned. In many cases, for instance, the land management agency should begin organizing the process even during exploration, when it becomes evident that the mineral resources appear sufficient to support mining.

Implementation

The land management agencies can promote the early consultation process by undertaking an aggressive, critical review of their current NEPA implementation practices and identifying sources of delay and inefficiency. Examples of effective and timely permitting also should be examined and used as models to improve and streamline the permitting of mining projects elsewhere.

In addition, it is crucial that the land management agencies adopt policies that will ensure that they secure high-level commitments to participate fully in the NEPA process from each potentially concerned federal, state, and local agency, and that will identify and include as cooperators the appropriate tribal and other nongovernment organizations. This must include full participation in the scoping process so that new issues are not introduced too late, and it must include concurrence on the appropriate work programs for baseline studies and other report components before the work is undertaken. Implementation of this recommendation also may require other federal agencies involved in decision making to adopt and implement guidance of their own to assure that the collaboration is early, effective, and sufficient. In this regard, the EPA Hardrock Mining Framework, if implemented, provides an example of such an agency commitment.[2]

---

[1] EPA, 1997a

[2] See draft EPA policy (EPA, 1997b).

The National Interagency Coordinating Committee formed in 1995 by interagency agreement among EPA, BLM, the Forest Service, and the National Park Service represents an approach to coordination among senior managers on a policy level that might also support implementation of this recommendation. However, the agreement will expire in 2000 unless it is extended, and its effectiveness depends on continued attention by senior managers.

Recommendation 11: BLM and the Forest Service should maintain a management information system that effectively tracks compliance with operating plans and environmental permits, and communicates this information to agency managers, the interested public, and other stakeholders.

Justification

The Committee was consistently frustrated by the inability of federal land management agencies to provide timely, accurate information regarding the status of mining projects under their jurisdiction. The lack of information appeared to be greatest among highly placed officials who have the greatest need to know. Consequently, those responsible for regulatory management and change, and for keeping the public and Congress adequately informed, appear to be severely limited in their ability to do so. Public testimony expressing concern that mining on federal lands is or is not adequately regulated was based mainly on anecdotal data and generally lacked systematic support.

Discussion

As stewards of the federal lands, federal land management agencies have a responsibility to track systematically the performance of mining projects under their jurisdiction and to communicate this information effectively to stakeholders. Although the Committee was told by field-level personnel that such tracking was generally accomplished, the systematic compilation of the information necessary to document this conclusion was found only in field office files.

This recommendation addresses two of the essential elements of a regulatory program (see Sidebar 3-1), a management information system and stakeholder communications. The objective is to improve the federal land management agencies' management capabilities and to increase the confidence of stakeholders that the agencies are adequately fulfilling their obligation to protect the environment at mining projects on federal lands.

Mining regulation depends on the application of a multitude of statutes and regulations by numerous agencies, which the Committee found, in the aggregate, to be generally effective in protecting the environment from the potential impacts of mining on federal lands. The Committee also was told that mine operators are comprehensively implementing permit requirements, conducting required monitoring, and appropriately reporting results.

It was not clear, however, that the federal land management agencies are systematically tracking operator compliance with the commitments made in permits, reviewing and evaluating monitoring results, and effectively communicating this information to agency officials and interested stakeholders. Information that could be usefully tracked, compiled, and reported includes:

- inspection reports of federal land managers and of other federal, state, and local agencies;
- summaries of monitoring data and reports that, in particular, characterize the degree of compliance with standards and criteria at mines and that highlight incidents of noncompliance;
- corrective actions taken to return sites to compliance;
- enforcement actions required to assure implementation of required corrective actions;
- status of reclamation at mine sites;
- status of financial assurances;
- financial assurance forfeitures required; and
- outstanding incidents of noncompliance requiring further action.

Compiled information should be summarized periodically and reported upward in the federal land management agencies. This same information should be readily accessible to the public, either as published reports or postings on the Internet.

Implementation

The federal land management agencies should review and compile compliance data from the management information systems used by cooperating federal, state, and local agencies to track and report compliance with their permits and environmental requirements. The agencies should not duplicate efforts. Other federal, state, and local agencies also have an obligation to maintain their own management information systems and to provide appropriate information to the land management agencies. Care in the design and implementation of an appropriate management information system can avoid excessive paperwork and diversion of personnel from field inspection duties.

Periodic public reporting is needed on the compliance status of individual mines, as well as aggregated district, state, and national compliance information. This would help allay public concerns that federal land management agencies are not adequately protecting the environment at mining operations on federal lands.

Recommendation 12: BLM and the Forest Service should carefully review the adequacy of the staff and other resources devoted to regulating mining operations on federal lands and, to the extent required, expand and/or reallocate existing staff, provide training to improve staff capabilities, secure supplemental technical support from inside and outside the agencies, and provide other support as necessary.

Justification

Some land management offices report that they have too few people to conduct inspections, review proposed operating plans, process appeals, and conduct other required activities. This concern extends beyond the simple numbers of people assigned to these functions. The responsibilities of the land management agencies involve many different scientific disciplines and can be technically very complex. Offices responsible for regulating mining projects may not always have access to the trained and experienced personnel required.

Discussion

The objective of this recommendation is to improve and supplement the technical capabilities of the land management agencies to increase efficiency and ensure environmental protection. Opportunities available to supplement the existing capabilities of land management agencies are various and include the following:

- expanding and/or reallocating existing staff;
- providing intra and inter-agency training to improve staff capabilities, including not only briefings and seminars but also rotational temporary assignments, field visits, and other means of exposing less-experienced staff to the accumulated knowledge of veteran personnel;

- securing supplemental technical support from within the agencies (e.g., the BLM Technical Services Center's support of hydrological analyses in Nevada) and from other public agencies (e.g., the U.S. Geological Survey's Humboldt River Basin study provides important technical support to a land management agency by assisting in the evaluation and management of cumulative hydrological impacts);
- developing and using centers of excellence where individuals with the appropriate training and experience are available to support a number of different field offices;
- providing field offices with sufficient contracting funds to retain outside experts as necessary;
- expanding the role of private sector registered professionals (e.g., in addition to designing engineered mine facilities such as tailings dams, leach pads, diversion structures, and waste dumps; specifying the monitoring necessary to determine whether the facility is operating appropriately; reviewing the resulting data periodically; and certifying that the facility is operating within its design parameters, and if necessary, specifying corrective actions);
- using tools that will increase staff productivity, such as conducting more frequent inspections of exploration and mining activities and using computer models to estimate reclamation costs, conduct viability analyses, and support other tasks; and
- providing necessary supplies.

Implementation

Several of these opportunities warrant early attention by the federal land management agencies. Staff training should be provided to improve staff capabilities. BLM should explore opportunities to expand the role of its Technical Services Center to assist in scoping and preparing environmental documents, specifying appropriate environmental monitoring, reviewing environmental monitoring data, and assisting on other technical matters. The Forest Service research stations may be able to provide similar assistance. BLM and the Forest Service should seek additional technical resources from the USGS and other appropriate agencies such as NSF, DOE/National Laboratories, EPA, and the Fish and Wildlife Service. In addition, the academic community could become better engaged in support of the land management agencies. The agencies also could expand the role of registered professionals from the private sector and increase the use of tools to enhance staff productivity.

Recommendation 13: BLM and the Forest Service should identify, regularly update, and make available to the public, information identifying those parts of federal lands that will require special consideration in land use decisions because of natural and cultural resources or special environmental sensitivities.

Justification

Lacking information on the important natural and cultural resources and environmental sensitivities of federal lands, miners are at risk of investing in exploration and then being denied a permit or being permitted under unfavorable terms. From the public's viewpoint, lack of prior environmental information may cause a mining project to advance to the point where environmental sensitivities are recognized too late and cannot be adequately accommodated. The best opportunity to avoid these unfortunate outcomes is through the prior identification of special environmental and resource situations. The public availability of this information will permit all stakeholders to make informed decisions regarding the use of federal lands.

Discussion

Riparian zones, cultural and religious sites, unique ecosystems, and other settings on federal lands are now generally recognized as deserving special consideration in determining the appropriateness and management of mining. In reserving selected areas of federal lands for special purposes, such as national parks or wilderness areas, Congress typically withdraws the land from competing uses, including mining. The Forest Service and BLM use administrative withdrawals within legislatively prescribed limits to preclude hardrock mining on federal lands for such facilities as offices and campgrounds and for protecting areas used for scientific studies, municipal watersheds, and similar public purposes. The Committee notes, however, that using withdrawal orders is a relatively inflexible tool for reflecting environmental sensitivities during permitting in site-specific situations.

The land use planning processes required for federal lands by the Federal Land Policy and Management Act (FLPMA) for BLM and the National Forest Management Act (NFMA) for the Forest Service provide for identification of land and resources deserving special environmental concern. FLPMA, for example, provides for the designation of "areas of critical environmental concern" (ACECs) on BLM lands where special management is needed "to protect and prevent irreparable damage to important historic, cultural, or scenic values, fish and wildlife resources or other natural systems or processes, or to protect life and safety from natural hazards" (FLPMA, Sec. 103). Areas defined in land use plans as requiring special protection can be withdrawn from mineral entry within legislatively prescribed limits.

The agencies' land use planning processes provide for public input and review. However, the time needed to prepare and revise plans limits their usefulness in providing timely information. In addition to the information on environmentally sensitive resources and conditions provided in the land use plans, additional information should be provided in advisories that identify the types of resources and areas that the agencies consider sensitive. This additional information could serve as an early signal for miners and others about conditions that are likely to require additional attention during the permitting process.

Implementation

BLM and the Forest Service should identify natural or cultural resources or environmental sensitivities on federal lands that require special consideration in land use planning, including that related to hardrock mining. The agencies should use their land use planning processes to (1) identify these lands that should be withdrawn from hardrock mining or may require special considerations in permitting, (2) give specific consideration to hardrock mining as a potential land use, and (3) establish guidelines for reclamation and mitigation that apply to mining. This can be accomplished through the land use plans for federal lands required by FLPMA and NFMA.

Adopting this recommendation should help to assure appropriate consideration of all surface resources and efficient use of resources and environmental information in mining decisions. When land use planning identifies lands that should be protected from all mining, existing withdrawal authority should be sufficient.

Recommendation 14: BLM and the Forest Service should plan for and assure the long-term post-closure management of mine sites on federal lands.

Justification

The Committee found that current regulatory programs primarily address the exploration, permitting, operation, reclamation, and closure of mines and have only recently focused on the post-closure management needs of mine sites on federal lands. Because hardrock mining on unpatented mining claims is a temporary use of the land and because the federal land management agencies resume direct responsibility for management of mine sites after reclamation and closure are complete and reclamation bonds are released, consideration should be given to the long-term needs and responsibilities for managing these sites.

Discussion

Federal land management agencies should consider the land uses that are appropriate for a closed and reclaimed mine site, and perhaps more importantly, whether any uses should be controlled or precluded. In addition, the following management requirements should be addressed and assured for each site:

- measures needed to preserve future mineral access;
- residual public safety hazards and the need for fences, signs, and other features that must be periodically checked and maintained;
- measures needed to assure the integrity of closed waste units including the monitoring of tailings pond caps and waste rock and leach pad covers and their possible repair because of erosion or other failure, and the checking of adit plugs for continued effectiveness;
- long-term environmental monitoring required to assure that the site remains stable and does not become a source of off-site contamination and the implementation of appropriate corrective measures;
- the operation and maintenance of any water treatment facilities required to maintain water quality compliance of the site over the long term; and
- financial assurance to ensure implementation of these post-closure management requirements.

These matters are more relevant to large sites than small, but should be considered on a site-specific basis for any operation subject to a plan of operations. Site-specific post-closure management plans should be prepared and approved, and to the extent that costs can be reasonably identified, appropriate post-closure financial assurances should be established for mine sites on public lands.

The need for appropriate consideration of sites reasonably anticipated to require long-term operation of water treatment facilities is of particular importance because of the substantial costs of operating and maintaining these facilities. Many questions remain about maintenance of treatment facilities for long periods of time and the types of financial assurances that might be appropriate to cover such costs. Questions also remain about the predictive tools currently available, their ability to accurately predict the need for long-term water treatment (see Appendixes B and D), and how and when to apply the results of such predictions to the need for financial assurances.

The Committee did not examine the long-term treatment and associated financial assurances in great detail. However, information was received indicating concern by the public and regulatory personnel about long-term treatment and the ability to require financial assurances. The Committee found that, with some exceptions, most states, the BLM, and the Forest Service do not require bonding for long-term or perpetual treatment of water at mine sites. The need for treatment of acid drainage and pit waters may extend long beyond typical post-closure periods (generally 30 years or less). If these costs are not assured, the public will eventually incur the costs of long-term water treatment at such sites if they become necessary. The Committee agreed that long-term water treatment and associated financial assurances will be necessary at some sites, and that the appropriate application of the predictive models and the appropriate long-term financial assurance instruments are areas that need further study.

The Committee reached no conclusion regarding the appropriateness of either requiring financial assurances for unanticipated, potentially environmentally damaging, extremely low-risk post-closure occurrences (such as the failure of drainage control systems or tailings embankments in a storm that is greater than the design storm event).

Implementation

The federal land management agencies should develop procedures that will enable them to identify, in the review and approval process for plans of operations, the kinds of post-mining requirements that are likely to arise and to incorporate these into the approved plan of operations. In addition, provisions should be made to amend or clarify, as necessary, applicable land use plans to reflect the post-closure requirements of the site and to consider institutional, management, staffing, and other needs of the post-closure mine site. Appropriate types of financial assurance should be investigated for long-term water treatment.

Recommendation 15: BLM should prepare guidance manuals and conduct staff training to communicate the agency's authority to protect valuable resources that may not be protected by other laws.

Justification

Although mining operations are regulated under a variety of environmental protection laws implemented by a number of different federal and state agencies, the Committee was concerned that these laws may not adequately protect all the valuable environmental resources that might exist at a particular location proposed for mining development. Examples of resources that may not be adequately protected include springs, seeps, riparian habitat, ephemeral streams, and certain types of wildlife. In such cases, the BLM must rely on its general authority under FLPMA and the 3809 regulations to prevent "unnecessary or undue degradation." Because the regulatory definition of "unnecessary or undue" at 3809.0-5(k) does not explicitly provide authority to protect such valuable resources, some of the BLM staff appear to be uncertain whether they can require such protection in plans of operations and permits. Some resources may deserve to be protected from all impacts, while other resources may withstand some impacts with associated mitigation. BLM should clarify for its staff the extent of its present authority to protect resources not protected by specific laws, such as the Endangered Species Act.

Discussion

As discussed in Chapter 3, BLM and the Forest Service have a variety of mechanisms available for withdrawing from mining or protecting particularly valuable or sensitive areas or resources in advance of any proposed development. To the extent that they can make use of these mechanisms, they will provide protection to such resources or areas at the same time that they provide advanced notice to potential mining operations about which areas and resources ought to be avoided or protected in the planning and design of these operations. However, it may not be possible to identify all such areas or resources in advance. In this case, the agencies will have to rely on their general authorities, implemented through the NEPA process, to provide such protection on a site-specific basis.

Representatives from the Forest Service seemed clear in their authority to provide such protection, but some BLM representatives expressed ambiguity about this issue. The ambiguity results from the definition of "unnecessary or undue" in the agency's regulations pertaining to hardrock mining. These regulations emphasize such factors as "reasonable mitigation measures," "prevention of a nuisance," and compliance with other explicit environmental protection statutes. They do not, however, provide explicit authority to protect valuable or sensitive resources that are not adequately protected by these other environmental protection statutes. Nevertheless, BLM has required such protection in some operating plans, and the Department of the Interior's Board on Land Appeals has upheld these requirements.

However, there is no clear guidance for agency personnel in the regulations or in the guidance documents on the extent of the agency's authority or on how these issues should be handled in the development and approval of operating plans for proposed mining operations.

Implementation

BLM should prepare guidance manuals to clarify its existing authority to protect resources and train staff accordingly. During the NEPA process for mine operating plans, BLM should consider unnecessary or undue degradation of those resources that are not protected by other laws, determine what authority it has to act, and take appropriate actions to provide the appropriate level of protection.

Recommendation 16: BLM and the Forest Service should plan for and implement a more timely permitting process, while still protecting the environment.

Justification

The permitting process is cumbersome, complex, and unpredictable because it requires cooperation among many stakeholders and compliance with dozens of regulations for a single mine. As a result, there is a tendency for the process to drag on for years, even a decade or more. This drains and diverts the resources of land management agencies that should be managing their full range of responsibilities. It is also burdensome to operators and does not provide the best environmental protection. The public, the land management agencies, and the permit applicants would all benefit if the permitting process were conducted more efficiently.

Discussion

NEPA reviews and permitting are complex and time consuming because of the wide range of environmental and other issues and the numerous stake-holders with diverse priorities. The collection of some baseline environmental data requires at least a full cycle of seasons, and sometimes longer. All deserve thorough consideration. At the same time, the review and permitting processes should be completed as soon as the work can be done properly, eliminating delays due to inadequate stakeholder cooperation, insufficient planning, or insufficient agency staffing. An efficient process will require full disclosure of information related to a proposed operation, full public access to and participation in the process, and full cooperation of all stakeholders and agencies interested or involved in the proposed operation.

The efficiency of NEPA review and permitting is in large part a management matter. The land management agency with lead responsibility should set and achieve deadlines and have sufficient qualified staff to do so. More timely permitting will free the agency staff to better address all their other environmental responsibilities. Recommendations in this report that support more efficient reviews and permitting include Recommendations 1, 2, 6, 10, 11, and 12.

Implementation

Several of the other recommendations in this report, when implemented, will help achieve the objectives of this recommendation. Information on the time involved in recent reviews and permitting need to be compiled and analyzed to identify and correct causes of any unnecessary delays. With these actions land management agencies can set and manage to achieve reasonable goals for completion of reviews and permits while continuing to protect the environment.

SUMMARY

The above conclusions and recommendations, taken together, summarize the Committee's views of the actions needed to coordinate federal and state mine reclamation, operations, and permitting requirements and programs. Some of these will require congressional action and some will require changes in federal regulations. Still others will require changes in the implementation of existing regulations and programs. Adopting these recommendations will improve environmental protection and reclamation of hardrock mining on federal lands, as well as the efficiency of the permitting process.

References

American Geological Institute. 1997. Dictionary of Mining, Mineral, and Related Terms, 2nd Edition. Ventura Publishing 3.0

Anderson, J. A. 1982. Gold—its history and role in the U.S. economy and the U.S. exploration program of Homestake Mining Company. Mining Congress Journal 68(1):51-58.

Azenha, M., M. T. Vasconcelos, and J. P. S. Cabral. 1995. Organic ligands reduce copper toxicity in *Pseudomonas syringae*. Environmental Toxicology and Chemistry 14:369-373.

Barringer, S. 1999. Mining Regulatory Programs in the Western United States: A Survey of State Laws and Regulations. Unpublished report prepared for the Precious Metals Producers; Singer, Brown, & Barringer, Nevada.

Beebe, R. R. 1995. Mining Technology Vision Panel. Unpublished report prepared for U.S. Department of Energy and the National Partnerships for Mining and Minerals Technology. Salt Lake City, Utah.

Beltman, D. J., W. H. Clements, J. Lipton, and D. Cacela. 1999. Benthic invertebrate metals exposure, accumulation, and community-level impacts downstream of a hard-rock mine site. Environmental Toxicology and Chemistry 18:299-307.

Bird, D. A. 1993. Geochemical modeling of mine pit water: An overview and application of computer codes. Masters thesis. University of Nevada, Reno. Also EPA report: EPA/530/R/95/012, PB95-191250.

Blanchard, C. L., and M. Stromberg. 1987. Acidic precipitation in southeastern Arizona: Sulfate, nitrate and trace-metal deposition. Atmospheric Environment 21:2375-2381.

BLM (Bureau of Land Management). 1986. Visual Resource Contrast Rating Handbook. BLM/H/8410/1. Washington, D.C.: U.S. Department of the Interior, BLM

BLM. 1988. National Environmental Policy Act Handbook. BLM/H/1790/1. Washington, D.C.: U.S. Department of the Interior, BLM.

BLM. 1992. Public Land Statistics—1991. BLM/SC/PT-92/011+1165. Volume 176. Washington, D.C.: U.S. Department of the Interior, BLM.

BLM. 1994. Public Land Statistics—1993. Volume 178. BLM/SC/ST-94/001+1165. Washington, D.C.: U.S. Department of the Interior, BLM.

BLM. 1996a. Cortez Pipeline Gold Deposit: Final Environmental Impact Statement. Volume I. Battle Mountain, Nev.: U.S. Department of the Interior, BLM.

BLM. 1996b. Final Environmental Impact Statement: Twin Creeks Mine. Winnemucca, Nev.: U.S. Department of the Interior, BLM.

BLM. 1997a. Ruby Hill Project: Final Environmental Impact Statement. Battle Mountain, Nev.: U.S. Department of the Interior, BLM.

BLM. 1997b. Final Environmental Impact Statement: Lisbon Valley Copper Project. Moab, Utah: U.S. Department of the Interior, BLM.

BLM. 1998. Final Environmental Impact Statement: Olinghouse Mine Project. Carson City, Nev.: U.S. Department of the Interior, BLM.

BLM. 1999a. Surface Management Regulations for Locatable Mineral Operations (43 CFR 3809): Draft Environmental Impact Statement. Washington, D.C.: U.S. Department of the Interior, BLM.

BLM. 1999b. Public Land Statistics—1998. Volume 183. BLM/BC/ST-99/001+1165. Washington, D.C.: U.S. Department of the Interior, BLM.

BLM/EPA (U.S. Environmental Protection Agency). 1998. Draft Environmental Impact Statement: The Proposed Yarnell Mining Project. Phoenix, Ariz.: U.S. Department of the Interior, BLM. San Francisco, Calif.: EPA, Region IX.

BLM/MDEQ (Montana Department of Environmental Quality). 1996. Final Environmental Impact Statement: Zortman and Landusky Mines—Reclamation Plan Modifications and Mine Life Extensions. Volume I. Lewiston, Mont.: U.S. Department of the Interior, BLM; MDEQ.

Bolis, C. L., A. Cambria, and M. Fama. 1984. Effects of acid stress on fish gills. Pp. 120-122 in Toxins, Drugs, and Pollutants in Marine Animals, C. L. Bolis, A. Cambria, and M. Fama, eds. Berlin: Springer Verlag.

Borgmann, U., and K. M. Ralph. 1984. Copper complexation and toxicity to freshwater zooplankton. Archives of Environmental Contamination and Toxicology 13:403-409.

Brattstrom, B. H., and M. C. Bondello. 1983. Effects of off-road vehicle noise on desert vertebrates. Pp. 167-206 in Environmental Effects of Off-road Vehicles: Impacts and Management in Arid Regions, R. H. Webb and H. G. Wilshire, eds. New York: Springer Verlag.

Brown, V. M., T. L. Shaw, and D. G. Shurben. 1974. Aspects of water quality and the toxicity of copper to rainbow trout. Water Research 8:797-803.

Chakoumakos, C., R. C. Russo, and R. V. Thurston 1979. Toxicity of copper to cutthroat trout (*Salmo clarki*) under different conditions of alkalinity, pH, and hardness. Environmental Science and Technology 13:213-219.

Clements, W. H. 1994. Benthic invertebrate community responses to heavy metals in the upper Arkansas River Basin, Colorado. Journal of the North American Benthological Society 13:30-44.

Clements, W. H., D. S. Cherry, and J. Cairns, Jr. 1988. The impact of heavy metals on macroinvertebrate communities: A comparison of observational and experimental results. Canadian Journal of Fisheries and Aquatic Sciences 45:2017-2025.

Coope, J. A. 1991. Carlin Trend exploration history: Discovery of the Carlin deposit. Nevada Bureau of Mines and Geology Special Publication 13.

DeLonay, A. J., E. E. Little, J. Lipton, D. Woodward, and J. Hansen. 1995. Avoidance response as evidence of injury: The use of behavioral testing in support of natural resource damage assessments. In Environmental Toxicology and Risk Assessment, T. W. LaPoint, F. T. Price, and E. E. Little, eds. Philadelphia: American Society for Testing and Materials.

Dobra, J. L. 1997. The U. S. Gold Industry 1996. Special Publication 21. Reno: Nevada Bureau of Mines and Geology.

Dobra, J. L. 1999. The U. S. Gold Industry 1998, Special Publication 25. Reno: Nevada Bureau of Mines and Geology.

Eary, L. E. In press. Geochemical and equilibrium trends in mine pit lakes. Applied Geochemistry.

EPA (U.S. Environmental Protection Agency). 1978. Compilation and Evaluation of Leaching Methods. EPA/600/2/78/095. Washington, D.C.: Office of Solid Waste.

EPA. 1985. The Gold Book: Quality Criteria for Water. EPA/44015/86/001. Office of Water Regulations and Standards.

EPA. 1993. Diffuse NORM Wastes and Waste Characterization and Preliminary Risk Assessment. RAE-9232/1-2. Contract # 68-D20-155 prepared for EPA, Office of Radiation and Indoor Air.

EPA. 1994. Acid Generation Prediction in Mining. Draft. Washington, D.C.: Office of Solid Waste.

EPA. 1996. 1995 Updates: Water Quality Criteria Documents for the Protection of Aquatic Life in Ambient Water. EPA/820/B/96/001. Washington, D.C.: EPA, Office of Water.

EPA. 1997a. EPA Office of Inspector General Audit Report—1997: EPA Could Do More To Help Minimize Hardrock Mining Liabilities. E1DMF6-08-0016-7100223. Washington, D.C.: EPA, Office of Inspector General

EPA. 1997b. EPA's National Hardrock Mining Framework. BLMWO/PL/00/001/3041. Washington, D.C.: Office of Solid Waste and Emergency Response; Office of Enforcement and Compliance Assurance; Office of Water.

Erickson, R. J., D. A. Benoit, V. R. Mattson, H. P. Nelson, and E. N. Leonard, 1996. The effects of water chemistry on the toxicity of copper to fathead minnows. Environmental Toxicology and Chemistry 15:181-193.

Forest Service. 1992. A Conceptual Waste Rock Sampling Program for Mines Operating in Metallic Sulfide Ores with a Potential for Acid Rock Drainage. Ogden, Utah: USDA, Forest Service.

Forest Service. 1998. Land Areas of the National Forest System: As of September 1997. FS-383. Washington, D.C.: USDA, Forest Service.

Forest Service/WSDE (Washington State Department of Ecology). 1997. Crown Jewel Mine: Final Environmental Impact Statement. Tonasket, Wash.: U.S. Department of Agriculture, Forest Service; WSDE.

Gabriel, I. E., and D. T. Patten. 1994. Distribution of copper smelter emissions in southeastern Arizona using honey mesquite as bioindicator. Water, Air, and Soil Pollution 72:67-87.

Gabriel, I. E., and D. T. Patten. 1995a. Changes in the inorganic element concentration spectrum of mesquite foliage during operational and non-operational periods of a copper smelter. Water, Air, and Soil Pollution 81:207-217.

Gabriel, I. E., and D. T. Patten 1995b. Establishing a standard Sonoran reference plant and its application in monitoring industrial and urban pollution throughout the Sonoran Desert. Environmental Monitoring and Assessment 36:27-43.

Galbraith, H., K. LeJeune, and J. Lipton. 1996. Metal and arsenic impacts to soils, vegetation communities, and wildlife habitat in southwest Montana uplands contaminated by smelter emissions: I. Field evaluation. Environmental Toxicology and Chemistry 14:1895-1903.

GAO. (Government Accounting Office) 1996. Land Ownership Information on the Acreage, Management, and Use of Federal and Other Lands (Letter Report, 03/13/96, GAO/RCED-96-40).

Gentry, D. W. 1998. Constrained prospects of mining's future. Mining Engineering: 85-94.

Gough, L., W. Day, J. Crick, B. Gamble, and M. Henning. 1997. Placer-gold Mining in Alaska—Cooperative Studies on the Effect of Suction Dredge Operations on the Fortymile River. U.S. Geological Survey Fact Sheet FS-155-97.

Guilbert and Park, 1986. The Geology of Ore Deposits. W.H. Freeman and Company. New York. 985 pp.

Hagler Bailly Consulting, Inc. 1996. Supplemental Injury Assessment Report: Clark Fork River NPL Sites Natural Resource Damage Assessment—Lethal Injuries to Snow Geese, Butte, Montana. Prepared for State of Montana Natural Resource Damage Litigation Program, Helena.

Howarth, R. S., and J. B. Sprague. 1978. Copper lethality to rainbow trout in waters of various hardness and pH. Water Research 12:455-462.

Interstate Mining Compact Commission. 1997. Noncoal Mineral Resource Report. Herndon, Virginia.

Kapustka, L., J. Lipton, H. Galbraith, D. Cacela, and K. LeJeune. 1996. Metal and arsenic impacts to soils, vegetation communities, and wildlife habitat in southwest Montana uplands contaminated by smelter emissions: II. Laboratory phytotoxicity studies. Environmental Toxicology and Chemistry 14:1905-1912.

Kiffney, P. M. and W. H. Clements. 1993. Bioaccumulation of heavy metals by benthic invertebrates at the Arkansas River, Colorado. Environmental Toxicology and Chemistry 12:1507-1517.

Kiffney, P. M., and W. H. Clements. 1996a. Size-dependent response of macroinvertebrates to metals in experimental streams. Environmental Toxicology and Chemistry 15:1352-1356.

Kiffney, P. M., and W. H. Clements. 1996b. Effects of heavy metals on stream macroinvertebrate assemblages from different elevations. Ecological Applications 6:472-481.

Kummerow, M. 1992. Weeds in wilderness: A threat to biodiversity. Western Wildlands 18:12-17.

Lauren, D. J., and D. G. MacDonald, 1986. Influence of water hardness, pH and alkalinity on the mechanisms of copper toxicity in juvenile rainbow trout, *Salmo gairdneri*. Canadian Journal of Fisheries and Aquatic Sciences 43:1488-1496.

LeJeune, K., H. Galbraith, J. Lipton, and L. A. Kapustka. 1996. Effects of metals and arsenic on riparian soils, vegetation communities, and wildlife habitat in southwest Montana. Ecotoxicology 5:297-312.

Leshy, J. D. 1987. The Mining Law: A Study in Perpetual Motion. Washington, D.C.: Resources for the Future, Inc.

MacDonald, M. S., G. C. Miller, and W. B. Lyons. 1994. Water Quality in Open Pit Precious Metal Mines. University of Nevada, Reno. Also EPA/530/R/95/011, PB95-191243.

MacRae, R. K., A. S. Maest, and J. S. Meyer. In press. Selection of an organic-acid analogue of dissolved organic matter for use in toxicity testing. Canadian Journal of Fisheries and Aquatic Sciences.

Marr, J., H. L. Bergman, J. Lipton, and C. Hogstrand. 1995a. Differences in relative sensitivity of naive and metals-acclimated brown and rainbow trout exposed to metals representative of the Clark Fork River, Montana. Canadian Journal of Fisheries and Aquatic Sciences 52:2016-2030.

Marr, J., H. L. Bergman, M. Parker, W. Erickson, D. Cacela, J. Lipton, and G. R. Phillips. 1995b. Relative sensitivity of brown and rainbow trout to pulsed exposures of an acutely lethal mixture of metals typical of the Clark Fork River, Montana. Canadian Journal of Fisheries and Aquatic Sciences 52:2005-2015.

Marr, J. C. A., J. Lipton, D. Cacela, J. A. Hansen, H. L. Bergman, J. S. Meyer, and C. Hogstrand. 1996. Relationship between copper exposure

duration, tissue copper concentration, and rainbow trout growth. Aquatic
Toxicology 36:17-30.

Marr, J. C., J. Lipton, D. Cacela, J. A. Hansen, J. S. Meyer, and H. L.
Bergman. 1999. Bioavailability and acute toxicity of copper to rainbow trout
(*Oncorhynchus mykiss*) in the presence of organic acids simulating natural
dissolved organic carbon. Canadian Journal of Fischeries and Aquatic
Sciences 56(8):1471-1483.

McElfish, J. M., T. Bernstein, S. P. Bass, and E. Sheldon. 1996. Hard Rock
Mining: State Approaches to Environmental Protection. Washington, D.C.:
Environmental Law Institute.

Nelson, R. L., M. L. McHenry, and W. W. Platts. 1991. Mining. American
Fisheries Society Special Publication 19:425-457.

Nelson, S. M., and R. A. Roline. 1993. Selection of the mayfly *Rhithrogena
hageni* as an indicator of metal pollution in the Upper Arkansas River.
Journal of Freshwater Ecology 8:111-119.

Nelson, S. M., and R. A. Roline. 1996. Recovery of a stream macroinvertebrate
community from mine drainage disturbance. Hydrobiologia 339:73-84.

Northwest Mining Association. 1997. Permitting Directory for Hard Rock
Mineral Exploration. Spokane, Washington.

NRC (National Research Council). 1979. Surface Mining of Non-Coal Minerals:
A Study of Mineral Mining From the Perspective of the Surface Mining
Control and Reclamation Act of 1977 (COSMAR Report). Washington,
D.C.: National Academy Press.

NRC. 1990. Competitiveness of the U.S. Minerals and Metals Industry.
Washington, D.C.: National Academy Press.

Oppenheimer, M., and C. B. Epstein. 1985. Acid deposition, smelter emissions
and the linearity issue in the western United States. Science 229:859.

Patten, D. T., J. C. Stromberg, and M. R. Sommerfeld. 1994. Water and
Riparian Resources of the Santa Cruz River Basin: Best Management
Practices for Water and Resource Quality. Final Report to Southwest Center
for Environmental Research and Policy. Salt Lake City: University of Utah.

Playle, R. C., D. G., Dixon, and K. Burnison. 1993. Copper and cadmium
binding to fish gills: Modification by dissolved organic carbon and synthetic
ligands. Canadian Journal of Fisheries and Aquatic Sciences 50:2667-2677.

Price J.G., Shevenell, L., Henry, C.D., Rigby, J.G., Christensen, L., Lechler,
P.J., Desilets, M., Fields, R., Driesner, D., Durbin, B., and Lombardo,
W., 1995. Water quality at inactive and abandoned mines in Nevada:
Nevada Bureau of Mines and Geology Open File Reports 95-4, 72.

Price, W.A. and Errington, J.C., 1998. Guidelines for Metal Leaching and Acid
Rock Drainage at Minesites in British Columbia. Minstry of Energy and
Mines, August.

Price, W.A. 1997. Draft Guidelines and Recommended Methods for the Predicition of Metal Leaching and Acid Rock Drainage at Minesites in British columbia. Ministry of Employment and Investment. Energy and Minerals Division.

Roscoe, W. E. 1971 Probability of an exploration discovery in Canada. Canadian Mining and Metallurgical Bulletin, Vol. 64, (707): 134-137.

Round Mountain Gold Corporation. 1994. Round Mountain Gold, An introduction for visitors. Unpublished company report.

Runnells, D. D., M. J. Shields, and R. L. Jones. 1997. Methodology for adequacy of sampling of mill tailings and mine waste rock. In Tailings and Mine Waste, A. A. Balkema, ed. Ft. Collins, CO: Rotterdam and Brookfield.

Schaefer, D. H., and J. R. Harrill. 1995. Simulated Effects of Proposed Ground-Water Pumping in 17 Basins of East-Central and Southern Nevada. USGS Water Resources Investigations Report 95-4173.

Schafer, W. M. 1993. Design of geochemical sampling programs. Presentation at Mine Operation and Closure Short Course. Helena, MT., April 27-29, 1993.

Scott, M. L., P. B. Shafroth, and G. T. Auble. In press. Responses of riparian cottonwoods to alluvial water declines. Environmental Management.

Shokes, T. E., and G. Moller. 1999. Removal of dissolved heavy metals from acid rock drainage using iron metal. Environmental Science and Technology 33:282-287.

Simberloff, D., and L. G. Abele. 1982. Refuge design and island biogeographic theory: Effects of fragmentation. American Naturalist 120:41-50.

Sorenson, E. M. 1991. Metal Poisoning in Fish. Boca Raton, Fla.: CRC Press, Inc.

Stromberg, J. C., and D. T. Patten. 1992. Response of *Salix lasiolepis* to augmented stream flows in the upper Owens River. Madrono 39:224-235.

Stromberg, J. C., J. A. Tress, S. D. Wilkins, and S. D. Clark. 1992. Response of velvet mesquite to ground water decline. Journal of Arid Environments 23:45-58.

Stromberg, J. C., R. Tiller, and B. Richter. 1996. Effects of groundwater decline on riparian vegetation of semiarid regions: The San Pedro, Arizona. Ecological Applications 6:113-131.

Sullivan, M. E. 1991. Heavy Metal Concentration in Riparian Vegetation Exposed to Wastewater Effluent. Masters thesis. Arizona State University, Tempe.

Swenson, R.W. 1968. Legal aspects of mineral resources exploitation. In History of Public Land Law Development, P.W. Gates, ed. Washington, D.C.: Government Printing Office.

Tempel, R. N., L. A. Shevenell, P. Lechler, and J. Price. In preparation. Geochemical modeling approach to predicting arsenic concentrations in a mine pit lake.

Tingley, J. V., and B. R. Berger. 1985. Lode Gold Deposits of Round Mountain, Nevada. Nevada Bureau of Mines and Geology Bulletin 100.

Tingley, J. V., and H. F. Bonham, Jr. 1998. Major precious-metal deposits. In The Nevada Mineral Industry 1997, J. G. Price, J. V. Tingley, D. D. LaPointe, H. F. Bonham, Jr., S. B. Castor, and D. A. Davis, eds. Nevada Bureau of Mines and Geology Special Publication MI-1997.

University of California. 1988. Mining Waste Study Final Report. Prepared by the Mining Waste Study Team of the University of Calfornia at Berkeley, July. 416 pp.

U.S. Geological Survey. 1999. Mineral Commodity Summaries 1999. Washington, D.C.

U.S. Public Land Law Review Commission. 1970. One Third of the Nation's Land. GPO, Washington D.C.: Government Printing Office

Wanty, R .B., B. Wang, and J. Vohden. 1997. Studies of Suction Dredge Gold-Placer Mining Operations Along the Fortymile River, Eastern Alaska. U.S. Geological Survey Fact Sheet FS-154-97.

Wood, C. W., and T. H. Nash III. 1976. Copper smelter effluent effects on Sonoran Desert vegetation. Ecology 57:1311-1316.

APPENDIXES

Appendix A

The Nature of Mining

The purpose of this Appendix is to provide an introduction and basic understanding of the nature of ore deposits and various mining activities. Some environmental issues are mentioned, but a detailed discussion of potential environmental impacts of mining activities is presented in Appendix B. Appendix D includes a discussion of the directions that mining will likely take in the future, along with identification of research needs for the industry, particularly regarding environmental issues.

DESCRIPTION OF ORE DEPOSITS

Ore deposits form as variants of such geologic processes as volcanism, weathering, and sedimentation operating with an extraordinary intensity. Ore deposits typically are parts of large-scale (several miles across and perhaps just as deep) ore-forming systems in which many elements, not just those of economic interest, have been enriched. For example, arsenic, antimony, thallium, and mercury are commonly enriched in or near Carlin-type gold deposits. Explorationists continually seek to discern trace chemical haloes or geophysical patterns to combine with geological observations and concepts to recognize faint clues to the location of the ore deposit. Known ores constitute less than 1 part in 10,000 of the metal endowment of the upper 1 km of continental crust; thus, by far the largest portion of metals resides in ordinary rocks as a low-level background geochemical signature in amounts too meager for economic mining. Most ore deposits that cropped out in the lower 48 states have been discovered; new discoveries come from unrecognized extensions of known deposits or from "blind" ore bodies buried under alluvium or sedimentary or volcanic rocks. The Twin Creeks deposit near Getchell, Nevada is a good example of such an ore body. Many discoveries of ore deposits are made in or near old mining districts, in part because the same hydrothermal system that formed the high-grade, easy-to-find deposits that were previously mined commonly formed other deposits nearby (either laterally or deeper). The Round Mountain deposit mentioned later is an excellent example.

Many hardrock commodities are associated with magmatic and hydrothermal processes (Guilbert and Park, 1986). These processes, in turn, are associated with modern or ancient mountain belts. Mountainous or sparsely vegetated terrains, such as those in the western states, expose possibly productive rocks much more fully than do, for example, the mid-continent prairies. In addition, the West is blessed with geologic conditions, including abundant igneous rocks and associated hydrothermal systems, that have led to the formation of ore deposits. Thus, the prime prospecting ground is in land that many people regard as valuable for aesthetic reasons, which creates potential for conflict among uses of the land. Nonetheless, society requires both a healthy environment and sources of materials, many of which can be supplied only by mining.

One of the common features of most ores and the reason underlying much environmental concern is the presence of large quantities of sulfur, usually as the mineral pyrite (FeS_2), which creates the potential for environmental problems caused by acid runoff (see Sidebar 3-1 and Appendix B). A few mineral types (e.g., chromium or aluminum and placer deposits of gold, titanium, or tin) contain little sulfur and constitute minimal geochemical environmental problems. Many sulfide-bearing deposits contain abundant carbonate minerals, which buffer acid produced from the oxidation of sulfides, and many sulfide deposits that have been thoroughly oxidized by natural weathering no longer pose an acid mine drainage threat.

Many hydrothermal ore deposits have haloes of low-grade mineralization surrounding the ore. With better technologies (or higher prices) low-grade waste rock can become ore. A good example is the Round Mountain mine in Nevada (Figure A-1), where early production (350,000 ounces of gold from 1906 to 1969) was from high-grade veins mostly mined underground. From 1977 through 1997 about 4 million additional ounces were produced from a large open pit exploiting relatively low-grade disseminated mineralization in the rock surrounding the high-grade veins mined long ago; the projected eventual output will be more than 11 million ounces of gold (Tingley and Bonham, 1998), worth close to $3 billion at the current price of approximately $250/ounce.

Heap leaching has made possible the economic recovery of gold from low-grade ores. In some of today's major mining operations, grades can be lower than 0.01 ounce of gold per short ton of ore (0.3 gram of gold per metric ton of ore). Typical grades for open-pit gold mines in Nevada are about 0.07 ounce of gold per short ton of ore; underground mining is more costly and requires ores of about 0.3 ounce of gold per short ton to be profitable (Tingley and Bonham, 1998).

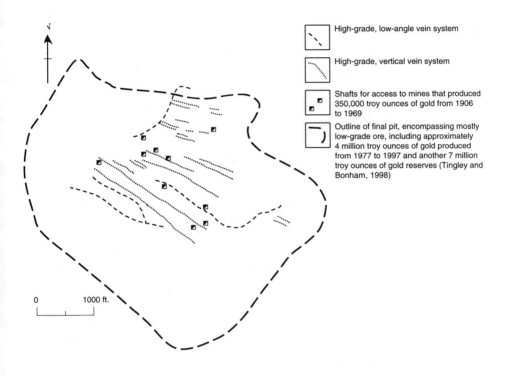

High-grade, low-angle vein system

High-grade, vertical vein system

Shafts for access to mines that produced 350,000 troy ounces of gold from 1906 to 1969

Outline of final pit, encompassing mostly low-grade ore, including approximately 4 million troy ounces of gold produced from 1977 to 1997 and another 7 million troy ounces of gold reserves (Tingley and Bonham, 1998)

0   1000 ft.

FIGURE A-1 Map view of the Round Mountain gold Mine. SOURCE: Tingley and Berger. 1985. Round Mountain Gold Corporation. 1994.

Some minerals seldom occur in sufficient quantities to be mined just for themselves. Instead they are recovered as by-products or co-products with other commodities. Examples of by-products are cadmium, indium, gallium, germanium, and thallium with lead-zinc ores and rhenium with copper-molybdenum ores. Coproducts (e.g., gold and copper in many porphyry copper deposits, lead and zinc in many deposits hosted in limestones, and silver and gold in many epithermal vein deposits) augment the value of each other and make deposits economic that might be unmineable for only one of them.

Most metallic mineral deposits that formed by magmatic or hydrothermal processes have complex geometries that influence the selection of mining method; in contrast, most economically viable coal deposits are relatively flat-lying sedimentary rocks. Some examples are illustrated in Figure A-2. As a result, metallic mineral deposits are typically mined in a downward or vertical fashion, while coal deposits are normally mined horizontally. As stated in the COSMAR report (NRC, 1979), backfilling of large open pits for hardrock mines is generally impractical, because the irregular lateral and vertical extents of the metallic deposits normally preclude them from being strip mined as simply as coal deposits, wherein overburden and waste rock can be recycled as backfill while mining proceeds from one area to the next.

THE MINING PROCESS

The mining process consists of exploration, mine development, mining (extraction), mineral processing (beneficiation), and reclamation (for closure). The first three steps in the mining process are characterized by dramatically increasing costs.

Exploration

The primary objective of exploration is to find an economic mineral deposit. The initial step in exploration is prospecting, which is the search for indications of a mineral deposit of potential significance. The objective is to define a target that suggests the occurrence of a mineral deposit worthy of subsequent testing with one or more exploration methods. These targets frequently result from the compilation and analysis of data germane to the type of mineral occurrence sought.

Target identification may result from relatively unobtrusive and proven approaches, such as detailed surface geologic mapping, sampling of outcrops, and observation of alteration patterns. More typically, modern prospecting embraces various types of geochemical sampling, geophysical techniques, satellite remote sensing, and other sophisticated methods for identifying the often subtle expressions of deeply buried mineral deposits.

In general, the environmental impact of prospecting is minimal. Surface access to or near the prospect is the primary requirement. When exploration on federal lands indicates a valuable deposit, the prospector acquires rights, if not previously completed, to develop it by staking claims and recording them with appropriate agencies. The area required for a large mine and its facilities

(1) STRATABOUND (Carlin, Deep West)

(2) STRUCTURAL "VEIN-LIKE" (Bootstrap-Capstone)

(3) STRUCTURAL STOCKWORK (Gold Quarry)

FIGURE A-2 Styles of mineralization of Carlin trend gold deposits. (Coope. 1991. p. 14).

(e.g., waste dumps, tailings ponds) is frequently a few thousand acres. Because of intermixed ownerships, this often involves a combination of federal and private lands for a single mine.

Exploration continues with target testing. Target exploration techniques vary; however, they usually focus on confirming the presence of a deposit and delineating its size, shape, and composition. In nearly every case, the primary exploration tool is drilling. Drilling into the subsurface is performed to determine both the lateral and vertical extent of the deposit, as well as other important deposit characteristics (e.g., continuity of mineralization, grade and associated trends, mineralogical relationships and characteristics, rock types, and local hydrologic information).

Surface disturbance resulting from exploration activities increases as these programs progress from preliminary to detailed. Most surface disturbance results from the construction of access roads and drill sites. The spacing or distance between drill sites is site-specific, but may be 100 feet or less in some geologic situations. Wider drill spacings may be acceptable for predicting mineral continuity and for deposits that are reasonably well understood and exhibit predictable results over greater intervals. Closer drill spacing is required for other types of deposits containing more variable or erratic mineralization. Surface disturbance also may result from trenching activities across zones of mineralization or from the collection of a larger bulk sample for metallurgical testing.

The subsurface environment also may be affected by the drilling program, although the effects are usually minor due to the use of practices such as sealing holes and the proper disposal of drill cuttings. A more significant subsurface impact is often that associated with the collection of a large bulk sample at depth for metallurgical testing.

In contrast with most other industries, hardrock mining has few alternatives relative to location, because economic occurrences of minerals are geologically and geographically scarce. Only a very small portion of Earth's continental areas, certainly less than .01%, contains the economic portion of its non-fuel mineral endowment. Thus, one cannot arbitrarily decide to build a mine here or there, but rather one must discover and mine those few places where nature has hidden its minerals (see Sidebar 1-4).

Over the last century and a half the western lands owned or formerly owned by the federal government have yielded metals worth hundreds of billions of dollars at today's prices. These lands continue to be major producers of gold (Figure A-3) and many other mineral resources. The U.S. Geological Survey estimates that many new deposits remain to be discovered (U.S. Geological Survey, 1999), but they will be deeper and more costly to find and extract.

Mine Development

When a deposit identified and defined from the exploration program has been judged economic and the required permits have been obtained, the next activity is mine development to prepare the deposit for extraction or mining. Infrastructure (e.g., power, roads, and water) is put in place throughout the mine and physical facilities. Planning and construction of support facilities (e.g., offices, maintenance shops, fuel bays, and materials handling systems) and the mineral processing facility is completed. Surface locations are delineated and prepared for waste or overburden material placement, heap leach piles (the chemical leaching of low-grade ores), stockpiles, tailings impoundments, and so on.

Near-surface deposits in open-pit mines are prepared initially for production by removing the overlying waste material or overburden for placement at surface waste dumps. These dumps are engineered structures that continue to grow both laterally and vertically until mining ceases; occasionally it may be possible to return waste to exhausted portions of the pit as mining progresses. Deeper deposits designed for underground mining are developed initially by gaining access to the mineralization through vertical or inclined shafts or horizontal adits. Underground drifts, crosscuts, raises, and ramps are excavated to provide the access needed to mine the ore.

At this stage in the mining process the nature of the environmental impacts is different and their magnitude is greater than those associated with exploration activities. The footprint of the mining operation is essentially defined at this point.

Mining (Extraction)

As mine development nears completion, there is a transition to the mining component, which refers to extraction of the in-place mineralized material, as well as associated waste rock as the mine becomes deeper and grows laterally.

Even though every ore deposit is different, generalizations can be made about the basic functioning of mining. Whether an ore deposit is mined by surface (open pit) or underground, most mines use the same basic operations:

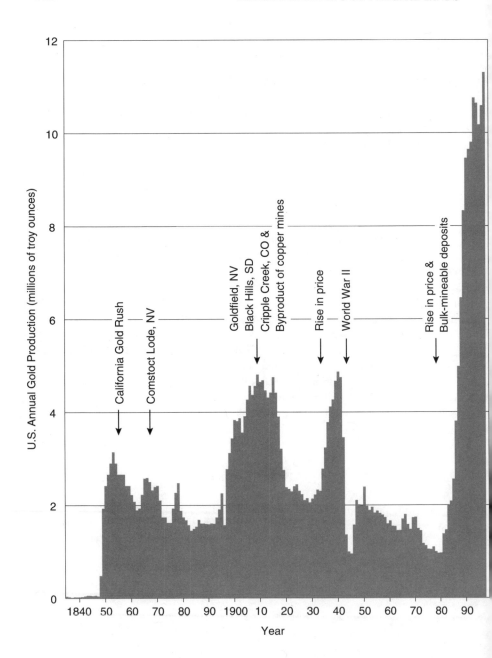

FIGURE A-3 Historic gold production in the United States Modified from Dobra, 1999.

drilling, blasting, mucking (loading), and transporting (hauling). At times a fifth operation of extending services, supplies, and ground control is needed.

Drilling and blasting refer to the drilling of holes for placement of explosive materials and detonation of the explosives contained in the drill holes. After blasting, the fragmented rock (muck) is typically loaded into some form of transportation system, in which it begins its journey to the mineral processing facility.

As mining progresses, open pits are excavated on the surface or a maze of underground openings or voids are created where the in-place ore was removed. In some cases, these relatively large underground voids (stopes) in the deposits may be backfilled with waste material, either for convenience or to enhance structural support, provide safety, or improve ore recovery.

Continued mining activities necessarily result in larger mines, along with growing waste dumps, heap leach piles, tailings ponds, etc. Many of the environmental impacts become obvious as one scans the surface; other impacts (i.e., impacts on the associated hydrologic regime) are more subtle and perhaps not so apparent.

Placer mining, in which gold is extracted from stream or beach sediments by gravity separation, is different from typical hardrock mining and mineral processing in several respects. Two types of placer mining are common on federal lands: (1) mining with mechanized earth-moving equipment and (2) suction dredging in streams. Placer mining with mechanized earth-moving equipment typically involves relocating a short (on the order of 2,000 feet) stretch of a stream, removal of the vegetative mat or soil, mining of gravels, removal of gold with sluices that separate dense from light minerals, and reclamation by replacement of gravel and the vegetative mat or soil.

Suction dredging is a technique that uses a pump to suction sediment from the stream bottom and process it in a floating sluice. Suction dredges can be small enough to be "recreational" one-person operations. States regulate mechanized placer mining both with suction dredges and with earth-moving equipment. The practice of hydraulic mining, wherein massive jets of water erode mineral-bearing gravels and wash them through an extraction facility, is practiced in a few places in the United States today, and these operations must comply with state and federal water quality discharge requirements.

Mineral Processing (Beneficiation)

Mineral processing, or beneficiation, consists of upgrading or concentrating the ore material before the concentrate is transported to a smelter or refinery. It begins by crushing and grinding the ore into small particles, thereby releasing interlocked ore grains from each other and from waste mineral grains. These particles are then subjected to various physical or chemical processes to separate and concentrate the valuable minerals from the unwanted waste and deleterious substances in the ore. Separation is designed specifically to take advantage of the unique physical or chemical properties of the valuable mineral(s). In many instances, various chemicals and reagents are used in the separation process.

Most notably with certain copper ores, metal can be produced on site directly through treatment of the ores by a process called solvent extraction/electrowinning. This process avoids the need to produce a concentrate prior to smelting.

The waste or unwanted minerals (tailings) separated from the valuable minerals, now in concentrated form, are routinely disposed of in a tailings pond near the mine site; the associated water is typically recycled, treated, and used in subsequent mining or processing activities. Tailings generally contain small amounts of the valuable mineral(s) not completely recovered during beneficiation; some unwanted or undesirable deleterious minerals; waste rock minerals; and some of the chemicals that may have been used in the course of the separation process. Underground mines may use these tailings to backfill mining-created underground voids.

Leaching is an increasingly common alternative extraction practice, almost universally applied (as cyanide solution) to the ore mined from some gold mines and (as ferric sulfate/sulfuric acid solution) to low-grade copper ore. In heap leaching, run-of-mine ore can be leached without crushing, or it is crushed only fine enough to allow the lixiviant (leaching solution) access to most of the mineral grains. The solution percolates through the piled ore, and the pregnant solutions are pumped to a processing facility for extraction. Then the fluids are treated and recycled. In some cases, uranium ores are leached *in situ* without mineralized material being removed or mined in a conventional manner (see following section on the uranium industry). While tailings dams and ponds and leach pads are carefully designed to high standards, the potential impacts resulting from release or discharge of tailings, leached rock, or pregnant leach solutions can be substantial.

Reclamation

Reclamation returns the mining and processing site to beneficial use after mining. A mine should not be closed until reclamation is complete, but as discussed extensively in this report, in some circumstances reclamation may never be fully accomplished, and long-term monitoring will be necessary.

Some common reclamation practices include lessening the slopes on the edges of waste rock dumps and heaps (to minimize erosion); capping these piles and tailings piles with soil; planting grasses or other plants that will benefit wildlife or grazing stock and help prevent erosion; directing water flow with French drains and other means to minimize the contact of meteoric water with potentially acid-generating sulfides in the waste rock dumps, heaps, and tailings piles; removing buildings; and eliminating roads to minimize unnecessary future entry by vehicles.

THE URANIUM INDUSTRY

The General Mining Law of 1872 classifies uranium as a hardrock mineral. Because special provisions apply to uranium, a brief discussion of regulatory issues related to uranium mining is given in this appendix. During the uranium "rush" of the 1950s and 1960s, a large number of mining claims were staked, and many uranium mines were developed on federal lands in Utah, Colorado, Wyoming, and other western states. However, because of the low price of uranium in the late 1990s, few uranium mines remain operating in the United States, primarily on BLM, private, and Indian lands. The need for uranium, principally as a fuel for commercial power plants that generate approximately 20% of U.S. electric power, is increasingly being met by imports. Uranium is also recovered as a by-product of other mineral mining activities, such as the production of phosphoric acid (from phosphate rock). In the United States uranium is also recovered from a variety of low-grade uranium-bearing rocks, from the mineral wastes of various processing facilities, and from groundwater that issues from an abandoned underground mine in New Mexico.

According to the EPA (1993), approximately 4 billion tons of mine-related wastes have been generated from surface and underground uranium mines. From the 1970s through the early 1990s open-pit mining produced almost 90% of that waste. According to the EPA, of the more than 1,300 open-pit uranium mines that have operated in the United States, most have been small, with only about 300 having a total ore production of more than 900 metric tons.

The Nuclear Regulatory Commission regulates uranium processing on public and private lands. It does not regulate uranium mining by traditional mechanical methods, but it does regulate *in situ* solution mining, because that extraction method is considered a form of processing. Traditional uranium ore mining is regulated by the usual federal and state agencies. Uranium mill tailings are regulated under the Uranium Mill Tailings Radiation Control Act (UMTRCA), which mandates special closure designs for uranium mill tailings ponds to prevent unacceptable release of radon gas to the environment.

Currently, there are only a few small underground uranium mines, and no active open-pit mines, operating in Colorado and Utah. Since the mid-1970s, *in situ* solution mining has been a commercially viable method for extracting uranium from subsurface ore bodies in porous sandstone. *In situ* uranium mining involves the injection through a borehole of an alkaline, oxidizing leaching agent (lixiviant) into the uranium-bearing sandstone at depth. The uranium is recovered, through production wells adjacent to the injection wells, from the pregnant lixiviant by passing the solution through ion-exchange columns; afterwards, the solution is chemically reconstituted and re-injected into the subsurface. The lixiviant consists of an aqueous solution of sodium carbonate-bicarbonate and/or carbon dioxide, and oxygen.

In situ mining occurs in Wyoming, Texas, and Nebraska. As mentioned earlier, the Nuclear Regulatory Commission is involved in the permitting and regulation of *in situ* uranium mines because the leaching of the ore is considered a form of chemical processing. The injection wells are considered by the EPA to be Class 3 underground injection wells, subject to regulation in the Underground Injection Control Program (40 CFR 143-147). Industry personnel report a significant degree of overlap between the regulatory requirements of the Nuclear Regulatory Commission and the EPA for injection wells. Several states have primacy over the Nuclear Regulatory Commission and the EPA for regulating radiation issues and underground injection.

The primary risk associated with *in situ* uranium mining is the potential for contamination of adjacent groundwater. If the system of injection and production wells is not properly designed and constructed, the pregnant lixiviant may escape into the sandstone aquifer, carrying with it dissolved uranium and radium. Small amounts of several trace metals are also present in the lixiviant, including lead, selenium, molybdenum, and arsenic. The requirements of the Underground Injection Control Program include restoration of the aquifer after completion of *in situ* mining. *In situ* mining also has been applied to other metals, especially copper, but not to the extent it has been used for uranium extraction.

Reclamation of uranium mines includes the typical steps considered at most other mine sites. Waste rock piles are regraded and revegetated, with the major objectives usually being control of erosion, physical stability, and restoration of the land for such uses as grazing and wildlife habitat. Open pits are evaluated relative to the physical stability of high walls and land use objectives, including pit lake water quality. Depending on the situation, the reclamation of a uranium mine may also include radiation-specific objectives. For example, reclamation of the large Jackpile mine in New Mexico used objectives for radiation control similar to the mill tailings program. At the Jackpile mine, a cover was constructed to limit the radiation exposure from the waste rock piles. Overburden enriched in uranium was placed in the bottom of the mine pit prior to backfilling. This selective burial of higher-grade overburden is consistent with reclamation practices used elsewhere, as in the

Wyoming and Texas abandoned mines reclamation programs (under the Wyoming Environmental Quality Act of 1973 and the Texas Surface Mining Act of 1975).

Most western states have regulations requiring the submission of a reclamation plan and bond prior to the issuance of a mining permit. Although these programs are not designed to mitigate potential radiological risks, such measures do reduce potential exposure to direct radiation, radon gas, fugitive dust emissions, and the possibility that waste materials will be removed and put to other uses, such as construction.

Many mines developed prior to implementation of regulations requiring reclamation are not yet reclaimed. Finally, there are examples of uranium mine reclamation in Superfund environments that follow the very stringent guidelines for reclamation specified in the UMTRCA.

Appendix B

Potential Environmental Impacts of Hardrock Mining

INTRODUCTION

From exploration through post-closure, hardrock mining has the potential to cause environmental impacts. In addition to the obvious disturbance of the land surface, mining may affect, to varying degrees, groundwater, surface water, aquatic and terrestrial vegetation, wildlife, soils, air, and cultural resources. Actions based on environmental regulations may avoid, limit, control, or offset many of these potential impacts, but mining will, to some degree, always alter landscapes and environmental resources. Regulations intended to control and manage these alterations of the landscape in an acceptable way are in place and are continually updated as new technologies are developed to improve mineral extraction, to reclaim mined lands, and to limit environmental impacts.

Some past mining practices have undeniably led to many of the potential impacts discussed in this appendix, and specific references are provided to describe the impacts in more detail where available. Although the committee was not successful in obtaining much information on recent environmental impacts at hardrock mine sites, the following three examples were obtained from a variety of sources, as listed below.

- Cyanide from a mine in Idaho was found to be contaminating a salmon stream in early 1999 (Idaho Division of Environmental Quality, May 21, 1999 News Release). The mine in question is currently in temporary shut down because the ore reserves were lower than expected. Cyanide was detected in routine quarterly monitoring conducted by company personnel, and the source of contamination is believed to be a lined tailings impoundment (personal communication, Nick Ceto, EPA Region 10).

- The Colorado Department of Public Health and Environment (CDPHE), Water Quality Control Division issued a notice of violation and cease and desist order for a gold mine that operated from 1989 until 1996 and is now in closure (notice of violation dated August 20, 1999, CDPHE).

The cause of the issuance was the discharge of manganese and sulfate that are leaching from backfill materials to a nearby creek through groundwater flow.

- The South Dakota Department of Environment and Natural Resources (DENR) issued a notice of violation and order for compliance to a gold mine for violation of the South Dakota Water Pollution Control Act (notice of violation, dated September 5, 1998, South Dakota DENR). The mine was found to be discharging effluent into streams in excess of permit levels for cadmium, copper, zinc, and total suspended solids, and its discharge failed a portion of the acute whole effluent toxicity test. Permit levels were exceeded once in 1996, twice in 1997, and nine times in 1998. In addition, the company failed to conduct daily monitoring of its effluent after the June 1998 violations and failed to promptly send samples for analysis, resulting in exceedance of holding times, both of which also constituted violations of its permit.

As discussed above, these examples are not necessarily representative of environmental impacts at all modern mines, but they do indicate that environmental impacts from some hardrock mining operations continue to exist under current regulatory conditions. However, the full extent of environmental problems at modern mine sites will not be known until better information on hardrock mine sites is collected and analyzed, as discussed in Chapter 3.

The Committee addresses the issue of mining-related impacts by presenting a discussion of the environmental impacts that have occurred, and that may still occur in some cases, even with regulations in place. It must be emphasized that these potential impacts will not necessarily occur, and when they do, they will not occur with the same intensity in all cases. Many of the impacts discussed in this appendix would violate current regulatory requirements and standards and would be subject to enforcement actions. Nevertheless, because some impacts continue, an understanding of the potential for mining to cause environmental impacts is essential to assessing and improving the regulation of hardrock mining on Federal lands.

CUMULATIVE EFFECTS

The environmental impacts of a single mining operation are broadly proportional to the size of the mine, although these vary depending on:

- the character of the mineral body and surrounding rock;
- the character of the environment directly and indirectly affected by mining;

- the character of the surrounding human environment;
- the nature of the mining operation; and
- the extent and effectiveness of actions to ameliorate the environmental impacts of mining.

Although a single mining operation creates its own set and degree of environmental impacts, a regional concentration of mines poses problems of cumulative impacts. For example, the number of gold mines along the Carlin Trend in Nevada and copper mines in the Miami and Globe area of Arizona poses environmental issues that can be addressed only if adequate consideration is given to their combined effects, along with those of other activities such as agriculture and timbering.

For example, groundwater withdrawal at a single mine has the potential to create a deep cone of depression in the local aquifer. As this cone expands over time, it may join those created by neighboring mines and lower the regional water table, which in turn may decrease or terminate flow in streams and springs some distance from the mines. Other regional uses of groundwater, such as agriculture and urbanization, contribute to the cumulative effects on the water table, but no one activity is fully responsible for this impact.

Similarly, potentially contaminated drainage or leachate from waste rock dumps, heap leach pads, or tailings ponds at a single mine may not be sufficient to lower stream water quality below acceptable concentration levels. But, when combined with discharges from neighboring mines and other contaminant sources, such as urban runoff and agriculture, the cumulative effects, especially in the semi-arid conditions of much of the West, has the potential to produce unacceptably high mass loadings.

In the Carlin Trend in Nevada, the Humboldt River receives discharged groundwater from the dewatering activites of several mines. Because the river ends in a closed-basin playa (Humboldt sink), the cumulative impacts of these discharges, along with highly contaminated irrigation return flows, have the potential to affect biota and water quality in this evaporative system. In addition, sediments and periphyton in the Humboldt River can concentrate constituents such as metals and metalloids over time and adversely affect macroinvertebrates. The USGS, the Forest Service, and other agencies are currently studying some of the cumulative impacts in the Humboldt River system.

Another example of cumulative impacts involves placer mining and suction dredging operations, which disturb streambeds and produce suspended sediments. The disturbed sediment usually settles out a short distance downstream if the activity is properly timed and controlled. However, there has been little evaluation of the cumulative impacts of sequential sediment dredges or placer mines.

Other issues involving cumulative impacts include the potential for contamination of groundwater aquifers, fugitive dust and air pollution from tailings and road surfaces, smelter emissions, and landscape degradation from large mine operations in the same general area. While most of the possible environmental effects of mining discussed in the following sections are in the context of individual mine operations, the potential for cumulative impacts also should be considered.

LONG-TERM MONITORING

Monitoring of environmental conditions and responses to human activities is needed to measure changes in the environment and to determine the effectiveness of mitigation procedures. There are different types of, or reasons for, monitoring (MacDonald et al., 1994):

- trend monitoring to measure changes in environmental parameters of interest (e.g., water quality components);
- baseline monitoring to characterize conditions prior to any action;
- implementation monitoring to assess whether an action was carried out as planned;
- effectiveness monitoring to determine whether a planned activity had the desired effect;
- project monitoring to assess the impact of an activity on an environmental component of concern (e.g., mining on water quality or stream biota);
- validation monitoring to validate particular predictive models used for planning (e.g., pit lake models); and
- compliance monitoring, to determine whether certain environmental criteria are being met (e.g., metal or pH standards in water).

Each of these types of monitoring may be applicable to the various stages of mining from exploration, through development and extraction, to closure, reclamation and post-closure. Baseline monitoring is essential to establish conditions prior to any mining activity and to provide the conditions against which future monitoring data can be compared. Monitoring during mining should address the implementation and effectiveness of environmental controls and compliance with regulations. Of equal concern is long-term monitoring to evaluate environmental protection during post-closure conditions and to validate predictive models.

Monitoring to address the effectiveness of mine closure activities and to validate predictive models related to closure must be long term for several reasons. First, conditions created during active mining operations might not lead to environmental impacts until after closure. This is especially true for water quality conditions, such as acid generation and migration of leachate from mine wastes to groundwater or surface water. Second, although design of post-closure conditions may have been based on best available information and technologies, unexpected events can alter future environmental conditions. The best available technologies may not have been sufficient to address, for example, long-term (decadal) changes in climatic patterns that may alter amounts and timing of precipitation events that influence the effectiveness of soil covers placed on heap leach pads and waste rock piles. In addition, post-closure use of these surfaces by recreationists, cattle, or other disturbances may be greater than expected, resulting in unanticipated erosion and releases from these closed areas. Third, predictive models used to design or plan closure procedures must be tested through monitoring to determine their accuracy. For example, the lack of robustness of pit lake models may lead to uncertainty about future conditions of surface and groundwater quality and quantity. Fourth, models designed to predict long-term responses of local and regional environmental parameters to post-closure conditions may be site specific and ignore the cumulative effects of other activities in the area, as discussed above.

Finally, monitoring to test predictions, calibrate models, and follow long-term trends of resources, such as groundwater levels and stream flows and chemistry, will produce information that is useful to researchers developing long-term predictive models. For this reason, long-term monitoring must be properly designed to measure those attributes that offer useful data on changes and/or sustainability of resources affected by mining activity and post-closure conditions.

WATER QUALITY

If not mitigated through regulation and prevention strategies, hardrock mining can have long-term impacts on water quality, which are defined in the Clean Water Act to include not only chemical but also biological and physical attributes. The water quality issues discussed below include metals and cyanide, acid drainage, pit lakes, and placer mining. Each of these issues also can have long-term impacts on water chemistry, aquatic biota, and aquatic habitat.

Metals and Cyanide

Hardrock mining of metalliferous deposits has the potential to release to the environment metals, metalloids (e.g., arsenic, antimony, selenium), sulfate, acid drainage, cyanide, nitrate, suspended solids, and other chemicals used in mining processes. These constituents are often present in waste rock, tailings, other mine waste materials, pit walls, and pregnant and barren solution ponds. Once the waste materials come in contact with water, the contaminants can be chemically leached and/or physically mobilized from the waste sources and transported downgradient to groundwater and surface water. Sediments and down-gradient soils may become contaminated and act as a secondary source of contamination to surface water and groundwater. Unless the waste materials are removed or effectively isolated from the environment by capping and/or lining, releases of contaminants, especially metals, can continue for long periods of time.

Some of these contaminants, such as cyanide in surface water and surface impoundments, nitrate, and sulfate, can be broken down or transformed into relatively innocuous constituents through microbial activity or photolytic reactions. The fate of cyanide and metal-cyanide complexes in groundwater is less well understood. The impact of metals and metalloids to down-gradient or downstream waters may be minimized by adsorption onto soils and sediment, precipitation of solid compounds, or dilution by groundwater or surface water. However, unlike many organic compounds, metals cannot be transformed to other compounds. Metals and metalloids can be transformed from one aqueous species to another or from one environmental medium to another, but metals are not lost through biogeochemical transformations. It is this lack of degradation that renders metals a long-term water quality concern.

Acid Drainage

Acid drainage that contains metals is another potential long-term water quality issue at some mine sites. The nature of the acid drainage reaction is such that it is difficult to stop the production of acid once it has begun. Lead mines that were operating at the time of the Roman Empire are still producing acid drainage 2,000 years later. In addition, acid drainage may take years to form or become a water quality concern. Although the factors that create acid drainage and that may minimize its impacts are well understood, there is little long-term monitoring data to use in predicting the extent of acid drainage at a given mine site. If a laboratory test shows that acid drainage will be produced, it is even more difficult to predict when it will start, how acidic it will be, or how high the metal concentrations will be. Predictability issues associated with acid drainage are discussed in the following section in Appendix D (see "Water Quality" in Appendix D).

Carbonate-bearing waste rock or tailings can produce drainage with an alkaline or near-neutral pH, but it will still contain elevated concentrations of components such as cadmium, zinc, manganese, arsenic, molybdenum, and selenium. Iron may or may not be elevated in circumneutral drainage water, depending on the oxidation state of the water.

Methods for prevention of acid drainage fall into two categories, those that prevent the acid generation from occurring and those that treat the acid generation at the source so that no drainage occurs. Although the latter is technically a treatment method, it stops the acid from draining and is different enough from the typical collect-and-treat methods to be considered here an acid drainage prevention method.

The most widely used methods for preventing acid generation include capping and sealing acid-generating rock to prevent air and water from reaching the rock. Such capping will require long-term maintenance to ensure that erosion, animal burrows, or other activities do not compromise the integrity of the cap. A variation on this method used in drier climates is to provide a less effective cap and a good vegetative cover, which will allow evapotranspiration of most of the water infiltrating into the pile. While not 100% effective in stopping water, such a cover may provide enough of a barrier to minimize the acid generation to levels that will not be detectable in ground or surface waters. Another variation on capping, which is practiced widely outside the United States, is burial of acid-generating materials in water to prevent the contact with air that is necessary to start the process. This is accomplished by placing the waste in a body of water, or by covering the top of a tailings pond with water once tailings deposition is completed. Some mines both in and outside the United States place potentially acid-generating materials into pits that are expected to fill with water. Once the pit lake is formed, the material no longer is exposed to air. To remain effective, water must always be over the material; this therefore requires long-term maintenance.

There are also specially developed chemical additives that, when applied to waste rock or spent ore piles prevent acid generation (for example, phosphate and surfactants.) Many of these additives are new and have not been studied for long-term effectiveness. These additives also can be expensive when used to cover a large area.

Certain types of recovery processes make it cost effective to separate the potentially acid-generating constituents in the ore from the rest of the ore. These methods concentrate the metals and sulfides, but render the majority of the ore benign and available for easy disposal. The concentrated material, which would have potentially more severe environmental impacts, can then be treated more cost effectively than the larger volume of spent ore material.

The most common method for treating in place to prevent acid drainage is to mix the acid-generating materials with a neutralizing material such as lime. The materials can be layered with lime or physically mixed together before or after disposal. The neutralizing material must be placed in ratios high enough to counteract all the acid-generating potential of the other material. The long-term effectiveness of this type of mixing is not known, and the relative rates of exhaustion of acid generation and neutralization have not been well studied.

Pit Lakes

Pit lakes have the potential to create long-term impacts on the environment that include major surface disturbances and alterations of pre-mining water quality and quantity. If water in a pit lake is contaminated and does flow to down-gradient groundwater and possibly to surface water, the impact of pit water on down-gradient waters will be another long-term water quality issue. Even if pit water does not flow down gradient, the concentration of metals, other contaminants, and salinity in the pit through evaporation may become a long-term water quality issue, especially for migratory birds and terrestrial wildlife. For example, waters of the Berkeley pit in Butte, Montana, were lethal to migrating snow geese that used the lake as a stopover in 1995 (Hagler Bailly Consulting, Inc., 1996). Most states are just beginning to establish water quality standards or designated uses for pit lakes. Standards that are established should be protective of designated uses, such as use by wildlife and migratory birds. The scientific basis for establishment of water quality standards for protection of wildlife and migratory birds needs more research. Predictability issues associated with pit lakes are discussed in Appendix D.

Surplus Water Discharge

Groundwater withdrawn to dewater pits is often discharged into local streams and drainages. These waters have the potential to be higher in total dissolved salts, metals, and other chemicals than the streams into which they are discharged. Although the normal flow of the stream may dilute these chemicals, they may become concentrated in stream sediments and periphyton, which can adversely affect some aquatic biota. Over the long term, this problem can become more serious if the stream ends in an evaporative sink that allows the contaminants to accumulate in the water column and sediments, which provide habitat and food for waterfowl and migratory birds.

Road Construction

Road construction during several phases of mineral exploration and extraction may disturb the soil surface sufficiently to create an excess of loose sediment that is carried off the site into local streams. Sediment may also be eroded from waste rock piles and may be carried by storm events down gradient into stream channels. The amount of sediment entering streams is a function of the quality of the road construction, the closeness of the streams, the types of hydrological events in the region that drive sediment transport, and the effectiveness of the erosion control measures employed.

Contamination from Railroads

Railroad beds constructed in mining areas may have been composed of mine wastes. This results in the placement of mine wastes over a greater area than normally anticipated in a mine operation. Railroads commonly have been located in valley bottoms where there is less topographic variability. Consequently, railroad beds historically have been placed adjacent to rivers. Storm events may leach chemicals out of the railroad beds into the streams or adjacent drainages. Occasional spills from rail cars carrying toxic materials may also contaminate surface waters.

WATER QUANTITY

Minewater Discharge to Streams

Surplus water accumulated from mine dewatering is, in some cases, discharged into nearby streams. The timing and amount of discharge has the potential to affect the down-gradient riverine ecosystem. Riparian vegetation has evolved to disperse seed, and recruit young plants on the declining limb of spring floods. If surplus water discharge occurs after this period, young plants established in the spring will be inundated and lost. Surplus flows also may erode or straighten stream channels, reducing availability of natural sandbar recruitment sites for riparian vegetation (Stromberg and Patten, 1992). Increased velocity of surplus flows also may dislodge aquatic biota such as attached macroinvertebrates, or disrupt fish habitat. For example, alteration of flows from mine water discharge could increase turbidity in streams and cause increases in embeddedness of gravels used for spawning.

Groundwater Withdrawal

Groundwater withdrawal for mineral processing and to prevent filling of open pits and underground excavations has the potential to affect local and regional groundwater quantities and levels. In the former case, much of the water used is kept on site, while in the latter the excess water may be temporarily stored in impoundments but eventually may be discharged into local streams. In both cases, groundwater withdrawal may affect the local water table. For example, groundwater withdrawn from the Santa Cruz River Basin in southern Arizona for mineral processing at a nearby copper mine is lowering the water table by many meters, and is drying up the river. The impact on riparian vegetation is significant (Patten et al., 1994). A groundwater pump test at several hundred feet of depth for a new copper mine in Arizona was shown to alter surface flows of a small perennial stream by 75 acre ft/year (a significant reduction for the size of the stream) through lowering of the shallow alluvial water table that supported surface flows (Carlota Mine EIS). This evidence of impacts of groundwater withdrawal on surface flows can be extended to the potential of surface flow declines resulting from groundwater withdrawal for pit dewatering. Of greater concern are the potential cumulative effects of dewatering wells and groundwater wells for processing associated with several neighboring mines. The cones of depression in the deep aquifer resulting from groundwater withdrawal may coalesce and affect regional spring and stream flows that are dependent on the aquifer. Models of potential effects of groundwater withdrawal from deep aquifers in Nevada show that the decline in the shallow basin-fill aquifer and regional water table may be delayed for several decades following initiation of groundwater pumping (Schaefer and Harrill, 1995). The delay is partly due to the time taken to dewater a large regional aquifer.

Pit Lake Interception of Groundwater

While groundwater is withdrawn to prevent filling of the mineral extraction pits, abandonment of the pits may create lakes that have the potential to affect the local shallow aquifer. If the pit is excavated on sloping terrain, the shallow aquifer may be intercepted on the upper edge of the pit and its water may drain. The shallow aquifer is interrupted and little or no flow will continue into the aquifer downslope of the pit depending on the local head of the deeper aquifer. Water input to the pit lake from the shallow aquifer may eventually interact with a deeper aquifer at lower levels in the lake. The hydrological balance among these entities, the pit lake and the shallow and deeper aquifer, is not well understood (see "Pit Lake Water Quality" in Appendix D).

Runoff from Hardened Surfaces

Construction of roads, parking lots, buildings, and the like reduces the amount of absorption surface in the mining area. Similar to an urbanized area, a mine area has the potential to produce more runoff to nearby streams than would be expected without alteration of the ground surface. The runoff may carry contaminants, for example, oils from vehicles and sediment from the roads or roadside areas, but perhaps the greatest concern is the acceleration of runoff during storms and a reduction in moisture percolating into the ground. This has the potential to affect stream hydrological functions, to change sediment transport characteristics, and to alter instream habitat for fish and other aquatic organisms.

AQUATIC BIOTA

Aquatic Biota Response to Metals and Cyanide

Most metals and cyanide are toxic to aquatic life (e.g., fish, macroinvertebrates) at low concentrations (Borgmann and Ralph, 1984; Sorenson, 1991; Marr et al., 1995a,b, 1996). Aquatic life criteria are designed to protect aquatic life from such impacts as death and reproductive and growth disorders. For example, at 100 mg/l hardness, the chronic (long-term) aquatic life criteria for dissolved cadmium and copper are 2.24 and 8.96 mg/l, respectively. At lower hardness values typical of mountain streams, the criteria are even lower. For example, at a hardness of 25 mg/l as $CaCO_3$, which is typical of a number of mountain streams, the chronic criteria for dissolved cadmium and copper are 0.18 and 2.74 mg/l, respectively (EPA, 1996). Very low concentrations of metals, even below the chronic criteria, may cause fish to avoid certain waters and impair their growth (DeLonay et al., 1995; Marr et al., 1996). This is an issue for anadromous (migrating from fresh to salt water) fishes such as certain threatened and endangered salmonids, which may avoid streams with low metal concentrations, resulting in the elimination of that species from the watershed. The presence of natural organic acid can decrease or delay the toxicity of metals to aquatic biota (Playle et al., 1993; Azenha et al., 1995; MacRae et al., in press; Marr et al., 1999). Although there are no specific aquatic life standards or criteria for aquatic sediments, macroinvertebrates, which serve as food for fish, live in sediments and eat periphyton (algae) that coat rocks in streams. Both sediments and periphyton often contain metals at concentrations many times higher than their concentration in overlying surface water. The impact of metals from mining activities on macroinvertebrates has been shown in a number of studies (Clements et al., 1988; Kiffney and Clements, 1993; Nelson and Roline, 1993; Clements, 1994; Kiffney and Clements, 1996a,b; Beltman et al., 1999). When water quality in streams containing acid drainage and metals from mining activity is improved, macroinvertebrate communities have been shown to recover (Nelson and Roline, 1996). Metal- or acid-rich groundwaters can also contaminate surface waters.

Aquatic Biota Response to Acid Drainage

If acid drainage enters surface water, metals in the drainage will have the same types of effects discussed above under metals and cyanide. Fresh water biota also have certain pH requirements, usually between pH 6.5 and 9 (EPA, 1985), and acid conditions can cause adverse conditions for fish (Bolis et al., 1984). Metals in acid drainage, after being diluted downstream by higher pH waters, can be precipitated out of solution and can coat streambed material with an iron-rich and heavy metal-rich cement. The cement can impair streambed habitat for fish and macroinvertebrates by physically embedding gravels. When the spaces between gravels are embedded with fine-grained sediment or floc, egg survival is threatened by a lack of oxygen. Higher hardness and alkalinity can decrease toxic responses of aquatic biota to metals (Brown et al., 1974; Howarth and Sprague, 1978; Chakoumakos et al., 1979; Lauren and MacDonald, 1986; Erickson et al., 1996). Mine drainage water may have higher hardness values than nearby surface waters because of increased leaching of silicates and other materials in the rocks, efflorescent salts, and calcium- and magnesium-containing solids, but the low pH values of acid drainage almost always result in lower alkalinity values.

Impact of Placer Mining and Suction Dredging on Aquatic Biota

Placer (mechanized) mining in active streams and suction dredge mining disturb to some degree streambed sediments, which provide habitat for macroinvertebrates and spawning habitat for salmonids (redds). Fine-grained sediment, if disturbed by placer operations, may move downstream and cause damage to spawning grounds for fish or to benthic biota. If flows are high enough, and there is sufficient sediment upstream, the streambed may return to near its original characteristics after springtime high flows. Depending on the degree of disturbance during low-precipitation years, the streambed may remain unsuitable for aquatic life habitat until high flows return it more to its original characteristics. The U.S. Geological Survey is investigating the environmental impact of suction dredge mining on turbidity and metal concentrations in an Alaskan river (Wanty et al., 1997; Gough et al., 1997). Studies on the impact of suction dredge and mechanized placer mining on macroinvertebrate and spawning habitat should also be conducted.

LANDSCAPE OR ECOSYSTEM ALTERATIONS

While many active mines have been developed around existing or historic mining sites, new mineral discoveries may also occur in remote, roadless, mountainous country where human activity has caused little disruption of these relatively pristine ecosystems. Early exploratory surveying by qualified geologists has no more impact on these remote areas than geological mapping or casual recreation. However, exploration activities designed to evaluate a mineral deposit have the potential to disrupt these remote locations. Development of exploration roads, use of mechanized equipment such as drill rigs or soft-wheeled off-road vehicles, or use of helicopters are all invasive activities in otherwise unaltered ecosystems.

TERRESTRIAL VEGETATION

General Disturbances

Any disturbance of a terrestrial ecosystem, whether natural or anthropogenic, results in a change in vegetational composition. Some ecosystems are disturbance systems requiring some form of natural disturbance to maintain their particular structure and composition. For example, chaparral ecosystems are fire disturbance systems, while most riparian ecosystems are flood disturbance systems. As a consequence of natural disturbance processes, these ecosystems are maintained in an early successional stage, while removal of the disturbances will allow succession to proceed to a later successional, non-disturbance plant community.

Development of roads or use of off-road vehicles, whether soft- or hard-wheeled, during exploration activities has the potential to disrupt the soil and low stratus vegetation, stimulating invasion by disturbed-site plant species (e.g., members of the mustard family). This may be little different than surface disturbance by burrowing animals, but the consequences of mining activities are potentially less localized.

Small alterations of topography by exploration activities may create new habitats for hydrophytes in low areas and xerophytes on elevated terrain. These terrain changes result in localized composition changes of plant species, which may in turn alter composition of local invertebrate and other associated species. In time, most of these small changes may, through biological and physical processes, return the vegetation to a state similar to predisturbance.

Smelter Emissions Fumigation

Plant communities adjacent to some historic mineral processing facilities have been altered by fumigation from smelter emissions (Gabriel and Patten, 1994). Uncontrolled smelters have released large amounts of sulfur dioxide (SO_2) that settled on the surrounding landscape. When combined with water in plant leaves, SO_2 is converted to sulfuric acid, which is often lethal to foliage and consequently the plant. Some plants are resistant to SO_2 fumigation, resulting in plant communities near smelters being composed of resistant plants and devoid of non-resistant plants. With increasing distance from the pollution source, the plant community composition gradually returns to normal (Wood and Nash, 1976). Where long-term smelter emissions historically affected surrounding forests and associated soils, for example near Anaconda, Montana, Coeur D'Alene, Idaho, and Salt Lake City, Utah, much of the surrounding hillsides still are devoid of trees. Apparently, present ecological conditions at these sites are such that forest recovery is either slow or prevented by changes in soils or microclimate.

Metal Toxicity

Contaminants in air pollution from mining processes have the potential to affect terrestrial vegetation. Metals from uncontrolled smelter emissions contaminate soils and then are taken up by plants, or the emissions may settle on foliage and be taken up directly into the leaves. Soils, vegetation, and wildlife habitat in areas affected by smelter emissions have been shown to contain elevated concentrations of metals, and vegetation abundance and species diversity have been severely affected (Galbraith et al., 1996; Kapustka et al., 1996; LeJeune et al., 1996). Depending on the concentration of pollutants influencing plant physiology, uptake of metals and other contaminants may or may not be lethal. For example, concentration of metals in foliage of vegetation near a smelter was reduced by 50% to 100% when the smelter was not operating, and yet total foliar necrosis occurred only immediately adjacent to the smelter (Gabriel and Patten, 1995a). Plants used as indicators of pollution near a smelter had foliage metal concentrations much greater than plants growing in a pristine unpolluted area (Gabriel and Patten, 1995b). Foliage sampled near the smelter had copper concentrations nearly 800% greater and arsenic about 500% greater than foliage sampled at a non-smelter site. Uptake of metal-contaminated water and sediments by plants also causes contaminated foliage and stems in riparian and wetland areas.

Exotic Plant Species

Selection of plant species for reclamation is based on suitability for future uses, such as grazing, resistance to abuse by future uses (e.g., overgrazing or recreational use), and tolerance of site conditions. Plant species that fit these selection criteria may be non-native species; therefore, reclamation procedures may increase the distribution of non-native species.

Vehicular traffic on and off mining sites also may disperse non-native plant species into active mining areas. Off-road vehicles used for exploration purposes may carry propagules of non-native species into relatively pristine backcountry areas. This type of introduction is not unique to mining exploration (Kummerow, 1992), but invasion of non-native species into the backcountry is enhanced through mining exploration, which may be the first mechanical intrusion into some of these areas.

RIPARIAN VEGETATION

Valley Fill

Riparian plant communities are found in valleys where stream flows and shallow groundwater support phreatophytes. Historically, some valleys have been used as areas for placing waste rock, leach pads, or tailings impoundments. Although there is greater awareness today of the ecological impacts of using valleys for deposits, valleys may, in some cases, still be used for the placement of mining facilities. Obviously, if a valley is filled, the vegetation in the valley will be destroyed. Once filled, the riparian vegetation that requires the conditions found at the bottom of the valley cannot be restored.

Altered Hydrology

Some mining activities have the potential to consume most of the locally available water through extensive groundwater withdrawal, which in turn may affect surface flows and shallow valley fill aquifers (Patten et al., 1994). Some mines may intercept the deep water table, potentially disrupting regional aquifers and reducing stream and spring flows (Nelson et al., 1991). Groundwater withdrawal can affect riparian vegetation some distance from a mine. Reduced flows and lowering of the alluvial aquifer directly affect phreatophytic riparian vegetation, which depends on this water source. A lowered water table will stress riparian vegetation, causing either mortality or reduced vigor (Stromberg et al., 1992, 1996; Scott et al., in press). Lowered shallow alluvial aquifers may not maintain riparian vegetation, resulting in replacement of riparian species with upland species.

Water Contaminants

Metal-contaminated water and sediments that reach wetlands or settle along streams in the wetland or riparian zone create a contaminated substrate for plants that take up the metals and store them in foliage and stems (Sullivan, 1991). Contaminated soils and sediments from mine sites have the potential to affect bed, bank, and floodplain sediments, as well as down-gradient riparian areas and wetlands some distance from the mine.

WETLANDS: HYDROLOGICAL AND SURFACE CHANGES

Wetlands, like riparian ecosystems, are dependent on a continuous supply of water. Any change in regional hydrology may affect wetlands, especially in the arid West. Many arid region wetlands develop at spring orifices. The wetlands and the spring pools often support threatened or endangered species (e.g., pupfish). In Nevada, for example, springs occur throughout the desert, where the deep regional aquifer supplies water to the surface through the basin fill aquifer. These spring wetlands may be very sensitive to changes in the hydrologic head of the regional or local aquifer resulting from groundwater withdrawal by mines, agriculture, or municipal water use. There is sufficient evidence to show that small changes in the hydrologic head may lower the water table several meters, resulting in the drying up of springs and associated wetlands (Schaefer and Harrill, 1995).

Wetlands also occur in depressions throughout much of the mountainous West. Off-road vehicles can greatly alter the stability of these wetlands by creating ruts that drain the water. Even use of soft-tired vehicles can produce linear depressions that create pools and tend to dry up the remaining wetland.

SOIL

Surface Alterations

The Earth's surface supports a wide variety of organisms, but it is also susceptible to ambient environmental changes such as drought and wet cycles. Such human activities as road building or construction, whether a result of mining or other processes, have the potential to greatly alter soil surfaces and affect soils to some depth. Commonly topsoil is set aside to be replaced with a hardened, durable road surface or some other facility. Topsoil may be used eventually for reclamation of the roaded area or in reclamation of other mining activities (e.g., topping for waste rock piles). Hardened road surfaces also cause runoff to the road's edge, increasing the potential for invasion of weeds. In cold climates, roads may be salted, which causes soil contamination along the roadbed. Temporary roads may sufficiently alter soil surfaces through changes in topsoil structure and chemistry to prevent short-term recovery following reclamation.

Erosion

Disturbance of soil surfaces by mining activities may leave soils susceptible to erosion if precautionary measures are not taken. Eroded soils have the potential to contribute to sediment output into local drainages causing reduction in water quality.

Air Deposition

Chemical particulates and metals from smelter emissions and blowing tailings have the potential to settle on soil surfaces near some mineral processing facilities. In such cases, contamination of soils decreases with distance from the contaminant source. In a study around a copper smelter in Arizona, several metals, including iron, manganese, and copper, were found to significantly decrease in concentration in soils with distance from the smelter (Gabriel and Patten, 1994). This was true for both surface soil and soil at 25 cm depth. Plants can accumulate contaminants found in soils and may pass these along to herbivores.

TERRESTRIAL WILDLIFE

Disruptive Activities

Exploration into relatively pristine ecosystems, although minimal in surface disturbance, has the potential to disrupt wildlife. The location of dispersed drill pads over an extended area may prevent wildlife from finding seclusion from this type of activity. Exploration roads and even tracks from soft-tire vehicles, can disrupt migration of small mammals and change behavioral patterns of larger animals. Permanent long-distance haul roads or railroads for mining purposes have the same potential impact on animal behavior as roads and railroads used for other purposes, that is, they may alter migration patterns by creating barriers and fragment animal territories (Simberloff and Abele, 1982). This is especially true for smaller animals.

Noise from vehicular traffic, including off-road vehicles, has the potential to disrupt wildlife (Brattstrom and Bondello, 1983), sometimes preventing normal reproductive processes because the noisy activity is located in or near a calving or whelping area for ungulates or canines, respectively. Noise from blasting also has the potential to disrupt wildlife. Although some wildlife may become accustomed to blasting noises, others will move from the area, potentially reducing the population of that species.

Consumption of Toxic Plants and Animals

Terrestrial wildlife, waterfowl, and migratory birds may consume plants and animals that have accumulated toxic materials. Plants growing in contaminated sediments accumulate metals in tissues (Sullivan, 1991). Bioconcentration in trophic dynamics of aquatic organisms in lakes and streams produces aquatic plants, macroinvertebrates, and fish with elevated levels of metals and other contaminants in their tissues. Consumption of these organisms by wildlife and birds continues the bioconcentration process, potentially creating toxic levels of metals and other chemicals in these organisms.

Wildlife Enhancement

Some environments created by mining activities have the potential to benefit wildlife. For example, abandoned mine tunnels may be used by bat communities, thus influencing closure procedures for these developments. Reclamation of waste rock sites and other surface disturbances may create extensive areas of forage that attract some species.

AIR QUALITY

Smelter Emissions

Emissions of particulates and sulfuric acid from smelters have the potential to produce extensive regional air pollution. For example, prior to the imposition of current air quality standards, several copper smelters located in central and southern Arizona had been shown to create an extended haze and potential acid deposition over much of southern Arizona under the right climatic conditions

(Blanchard and Stromberg, 1987). Smelter emissions from the Sonora region of Mexico have been linked to wet sulfuric deposition and acidification of lakes in the Colorado Rockies (Oppenheimer and Epstein, 1985). Changes in lake acidity can cause changes in aquatic biotic composition and chemical processes.

Fugitive Dust

Unpaved roads, tailings ponds, and other disturbed areas have the potential to become sources of fugitive dust if they are not kept damp, adequately revegetated, or otherwise controlled. Mining, loading, dumping, and crushing activities are also potential sources of fugitive dust as are the surfaces of some tailings deposits.

NOISE

Noise can play an important role in human as well as animal behavior. Constant noise from heavy equipment operations at mines adjacent to residential communities may cause persons who are sensitive to noise pollution to move away or change their behavior to try to avoid the noise.

Appendix C

Existing Regulatory Requirements

Hardrock mining and exploration on federal lands are subject to many federal and state permitting, operating, and reclamation requirements. The committee has not attempted to produce an independent catalogue of the regulatory requirements. It has, however, reviewed several recent compilations, such as those listed in Sidebar C-1.

In addition, the committee has compiled excerpts from recent environmental impact statements that list regulatory requirements for proposed mines. The EISs contain this information because the Council on Environmental Quality's (CEQ) regulations implementing NEPA require that a federal agency identify and list the federal permits or licenses required to implement the proposal being studied:

> The draft environmental impact statement shall list all federal permits, licenses, and other entitlements which must be obtained in implementing the proposal. If it is uncertain whether a Federal permit, license, or other entitlement is necessary, the draft environmental impact statement shall so indicate. (40 CFR 1502.25[b])

Although the CEQ regulation addresses only federal requirements, the Bureau of Land Management's National Environmental Policy Act Handbook (BLM, 1988) has a provision that expands the requirement to state and local permits:

> Authorizing Actions (40 CFR 1502.25[b]). Identify the actions necessary to authorize the proposed action or alternatives. Include Bureau actions to authorize applicant proposals (e.g., granting right-of-way) and actions by other Federal, State, and local entities which

would be required for implementation (e.g., obtaining State or local permits). (BLM NEPA Handbook, Section V.C.3.e[5], at p. V-17)

Tables C-1 through C-9 include excerpts from nine recent EISs that identify permits, licenses, and other regulatory requirements for recent mines in six states:

- Table C-1, Cortez Pipeline Gold Deposit, Nevada, January 1996
- Table C-2 Zortman and Landusky Mine Extensions, Montana, March 1996
- Table C-3, Twin Creeks Mine Project, Nevada, December 1996
- Table C-4, Crown Jewel Mine, Washington, January 1997
- Table C-5, Ruby Hill Project, Nevada, January 1997
- Table C-6, Lisbon Valley Copper, Utah, February 1997
- Table C-7, Olinghouse Mine, Nevada, February 1998
- Table C-8, Yarnell Mining Project, Arizona, June 1998: Regulatory Compliance Summary
- Table C-9, Yarnell Mining Project, Arizona, June 1998: Significant Issues Raised During Scoping

---

**SIDEBAR C-1 Recent Compilations of
Hardrock Mining Regulatory Requirements**

• Barringer, S. 1999. *Mining Regulatory Programs in the Western United States: A Survey of State Laws and Regulations.* Unpublished, 87 pp.

• Bureau of Land Management. 1999a. *Surface Management Regulations for Locatable Mineral Operations: Draft Environmental Impact Statement.*

• Environmental Protection Agency. 1997b. *National Hardrock Mining Framework.* Mine Waste Task Force.

• Interstate Mining Compact Commission. 1997. *Noncoal Mineral Resource Report.*

• McElfish, J M., T. Bernstein, S. P. Bass, and E. Sheldon. 1996. *Hard Rock Mining: State Approaches to Environmental Protection.* Washington, D.C.: Environmental Law Institute, 358 pp.

• Northwest Mining Association. 1997. *Permitting Directory for Hard Rock Mineral Exploration.* Spokane, Washington.

TABLE C-1 Cortez Pipeline Gold Deposit, Nevada, January 1996

Agency	Permit/Approval	Facet of Project	Time Requirements	Comments
FEDERAL				
U.S. Bureau of Land Management (BLM)	(1) Approval of Plan of Operations (Requires Environmental Assessment [EA] or EIS)	All activities on unpatented mining claims or involving right-of-way on federal land	120-180 days	If EIS required, could take 365 days or longer. Public notice required.
	(2) Right-of-Way Permits	Road and Powerline access on federal land	30-180 days	Data in Plan of Operations and EA can be used for this application.
U.S. Environmental Protection Agency (EPA)	Review of State Water and Air Permits	Surface and groundwater discharge permits; Air Quality Permit	30-60 days review	Review capacity.
U.S. Army Corps of Engineers	Section 404 Permit	Any filling or dredging of wetland/riparian areas	90-180 days	If nationwide permit acceptable. Individual permit could take 365 days or longer.
STATE				
Nevada Division of Environmental Protection (NDEP)	Air Quality Permits			
	(1) Air Quality Permit to Construct (ATC)	All aspects, including construction, that produce air contaminants, i.e., particulates, hydrocarbons, sulfur dioxide, etc.	Up to 95 days	Requires plans and specifications for air pollution control facility. Public notice/hearing required.

TABLE C-1, Continued

Agency	Permit/Approval	Facet of Project	Time Requirements	Comments
	(2) Air Quality Permit to Operate (ATO)	All aspects, including construction, that produce air contaminants, i.e., particulates, hydrocarbons, sulfur dioxide, etc.	180 days to demonstrate compliance after start-up	No public notice required.
	Water Quality Permits			
	(1) Water Pollution Control Discharge Permit	Mine, tailings, heap leach operation; review discharge and seepage potential.	Minimum of 165 days	Review of geotechnical design criteria to verify zero-discharge operation; Public notice required.
	(2) National Pollution Discharge Elimination System (NPDES)	Any discharge of wastewater to surface water (i.e., sediment control facilities)	180 days	Review of geotechnical design; Public notice required.
	(3) NPDES – Stormwater	Coverage by general mining storm water permit at each site	30 days	Review of site plan; No public notice required.
	Solid Waste Disposal	Disposal of solid, non-toxic waste, i.e., garbage, construction waste, etc.	14-90 days prior to construction	Site location, design, and operation plan.
	Reclamation Permit for a Mining Operation	Any surface-disturbing aspect of the project		

TABLE C-1, Continued

Agency	Permit/Approval	Facet of Project	Time Requirements	Comments
Nevada Division of Water Resources (NDWR)	(1) Permit to Appropriate the Public Water	Use of surface and groundwater	90-180 days; prior to construction	Requires data regarding source of water and annual consumption; public notice required.
	(2) Permit to Construct Tailings Dam	Any tailings dam over 10 feet high or impounding more than 10 acre-feet	45-120 days; prior to construction	Review of geotechnical design; no public notice required.
Nevada Division of Wildlife (NDOW)	(1) Industrial Artificial Pond Permit	All facets	To be determined	Regulate wildlife impacts and cyanide-related impacts.
	(2) Dredging Permit	Removal of material from or placing material in wetlands	10 days; prior to operation	In conjunction with Army Corps of Engineers.
Nevada Division of Health/Consumer Protection Services	(1) Sewage Disposal Plans	Sewage system plans	5-30 days; prior to construction	No public notice required.
	(2) Drinking Water Supply	Drinking water supply plans	5-30 days; prior to construction	No public notice required.
Nevada Division of Historic Preservation	Review project to determine impact on cultural resources	All surface disturbances	30-90 days; prior to operation	Submit legal description with map.
State Inspection of Mines	Notification of Opening or Closing of Mines	Mining	60-120 days	

TABLE C-1, Continued

Agency	Permit/Approval	Facet of Project	Time Requirements	Comments
COUNTY				
Planning Commission (varies from County to County)	Special Use Permit	All surface disturbances	60-120 days	Application should include detailed plan of operations.
	Building Permit	Surface facilities	30-60 days	Must have prior approval from Nevada State Health Division.

SOURCE: Final EIS, Cortez Pipeline Gold Deposit, Nevada (BLM, 1996).

TABLE C-2 Zortman and Landusky Mine Extensions, Montana, March 1996

Agency	Permit, License, or Review	Authority	Purpose/Status
Montana Department of Environmental Quality	• State Operational Permit (Metal Mine Reclamation Act)	Title 82, Chapter 4, Part 3, *et seq.*, MCA; ARM §26.4.101 *et seq.*	To allow mining while adequately providing for the subsequent beneficial use of the lands to be reclaimed. Approval is documented in a Record of Decision.
	- Reclamation Bond	§82-4-338, MCA	Required of Proponent to assure sufficient reclamation funding is available at mine closure or abandonment.
	- Monitoring Plans	§82-4-335(4)m, MCA	To assure compliance with state and federal environmental resource standards and criteria;coordinate with other governmental agencies.
	• Open Cut Permit	§82-4-442 *et seq.*, MCA	For excavation of shale and clay pits; includes reclamation plan and bond.
	• Montana Pollutant Discharge Elimination System Permit (MPDES)	ARM 16.20.1301 *et seq.*	To control discharge (including stormwater runoff) to surface waters by setting water quality limitations and requiring self-monitoring. Conditions for MPDES permits for Zortman and Landusky mines are under negotiation in 1996.
	• 401 Certification	Sec. 401, Federal Clean Water Act (33 USC 1341); Montana ARM 16.20.1701 *et seq.*	Require prior to the U.S. Corps of Engineers being able to issue a 404 Permit; and is applicable to all federal activities which results in a discharge to state waters.
	• Permit for Construction and Operation of Air Contaminant Source	Montana Clean Air Act, ARM §16.8.11 *et seq.*	To control emissions of more than 25 tons per year of particulate matter.

TABLE C-2, Continued

Agency	Permit, License, or Review	Authority	Purpose/Status
U.S. Bureau of Land Management (2)	• Approved Plan of Operations	43 CFR §3809	To allow for mineral exploration and development on U.S. lands administered by BLM. Approval incorporates management requirements to minimize or eliminate effects on other BLM resources. Approval is documented in a Record of Decision.
	- Monitoring Plans	43 CFR §3809	To assure compliance with state and federal environmental resource standards and criteria; coordinate with other governmental agencies.
	- Reclamation Plan	43 CFR §3809; BLM Solid Minerals Reclamation Handbook No. H-3042-1	In coordination with DEQ, to ensure all reclamation activities meet the guidelines in the Resource Management Plan and BLM Manual Section 3042.
	• Material Site Permit	43 CFR §3610; Mineral Material Regs.	Material sale contract to establish fair market value and reclamation procedures for limestone from proposed quarry.
U.S. Environmental Protection Agency	• Review and approve authority for various programs, including 404 permit	Section 309 of Clean Air Act; Clean Water Act; other environmental statutes	Various NEPA review, environmental enforcement and oversight authorities.
U.S. Army Corps of Engineers	• Section 404 Permit for placement of fill or dredge materials in wetlands or water of the U.S.	Section 404 of the Clean Water Act	To control discharge of dredge or fill material into waters or wetlands of the United States; including intermittent streams where a bed and bank are recognizable.

TABLE C-2, Continued

Agency	Permit, License, or Review	Authority	Purpose/Status
U.S. Fish and Wildlife Service	• Biological Assessment	Section 7, Endangered Species Act, Migratory Bird Act	For Endangered Species Act compliance. If it is determined that adverse effects would occur to threatened or endangered species as a result of the Zortman and Landusky Mine extensions, the lead agencies would consult with USFWS to determine if measures could be developed to protect the affected species.
Montana State Historic Preservation Office	• Review of project for compliance with regulations governing protection of cultural and historic resources	Section 106 of the National Historic Preservation Act; 36 CFR Part 800	If historical, archaeological, or other cultural resources are located in the project area, the State Historic Preservation Officer would advise the lead agencies on impact mitigation of sites eligible for nomination to the National Register of Historic Places.
Montana Department of Natural Resources and Conservation	• Water Rights Permit	Montana Water Use Act, Title 85, Chapter 2, MCA	Required if the Proposed Action would use or extract, through surface water diversion or groundwater withdrawal, state water in an amount exceeding 100 gallons per minute.
Phillips County	• Floodplain Development Permit	Title 76, Chapter 6, Part 113, MCA	Required for construction of facilities within designated 100-year floodplains.
Montana Hard Rock Mining Impact Board and "Affected" Local Government Units	• Fiscal Impact Plan	Hard Rock Mining Impact Act: Title 90, Chapter 6, Parts 3-4, MCA	To mitigate fiscal impacts on local government services; (not required for ZMI extensions).
Phillips County Conservation District	• 310 Permit	Natural Streambed and Land Preservation Act, Title 87, Chapter 5, Sections 501-509, MCA	For any activity that physically alters the bed or banks of a stream. MDFWP provides recommendations and consultation.

SOURCE: Final EIS Zortman and Landusky Mines, Reclamation Plan Modifications and Mine Life Extensions, Montana (BLM/MDEQ, 1996).

NOTE: (1) It is the responsibility of the operator to have knowledge of, and obtain, any federal, state, or local permits, licenses, approvals, or reviews required by such entities for construction, operation, or closure of these projects. This table does not present an exhaustive list; several other actions will likely be required for project permitting. Expected hazardous materials for the Zortman/Landusky sites, including provisions for spills, are detailed in Sections 3.14 and 4.14.

(2) BLM, as the lead federal agency in the mine permitting action, and the steward for the federal lands impacted by the mining activities, would be responsible for ensuring that permitted actions comply with a number of federal statutes and regulations implementing those laws, including: the American Indian Religious Freedom Act (Public Law 95-341), the Archaeological Resource Protection Act (Public Law 96-95); the National Historic Preservation Act (Public Law 89-665); the Archaeological and Historic Preservation Act (Public Law 93-291); and the Endangered Species Act (Public Law 93-205).

TABLE C-3 Twin Creeks Mine Project, Nevada, December 1996

Permit or Approval	Regulatory Agency
Plan of Operations and Mine Reclamation Permit Approval	U.S. Department of the Interior, Bureau of Land Management; Nevada Department of Conservation and Natural Reosurces, Division of Environmental Protection
Right-of-Way Permits	U.S. Department of the Interior, Bureau of Land Management
Review of EIS and Air Permits	U.S. Environmental Protection Agency
Section 404 Permit (Predischarge Notice)	U.S. Army Corps of Engineers
Artifical Pond Permit	Nevada Department of Conservation and Natural Resources, Division of Wildlife
Air Quality Surface Disturbance Permit; Air Quality Operating Permit	Nevada Department of Conservation and Natural Resources, Division of Environmental Protection
Water Pollution Control Permit; Renewal of National Pollutant Discharge Elimination System Permit; General Stormwater Discharge Permit; and Groundwater Protection Permit	Nevada Department of Conservation and Natural Resources, Division of Environmental Protection
Permit for Dam Construction; Permit to Appropriate Public Waters	Nevada Department of Conservation and Natural Resources, Division of Water Resources
Review Project to Determine Impact on Cultural Resources	Nevada Division of Historic Preservation
Special Use Permit	Humboldt County

SOURCE: Final EIS, Twin Creeks Mine, Nevada (BLM, 1996b).

TABLE C-4 Crown Jewel Mine, Washington, January 1997

FEDERAL GOVERNMENT

Forest Service
- Plan of Operations
- Special Use Permits (Right-of-Way, etc.)

Bureau of Land Management
- Plan of Operations
- Special Use Permits (Right-of-Way, etc.)

U.S. Army Corps of Engineers
- Section 404 Permit – Federal Clean Water Act (Dredge and Fill)

Environmental Protection Agency
- Spill Prevention Control and Countermeasure (SPCC) Plan
- Review of Section 404 Permit
- Notification of Hazardous Waste Activity (1)

U.S. Fish and Wildlife Service
- Threatened and Endangered Species Consultation (Section 7 Consultation)

Federal Communications Commission
- Radio Authorizations

Treasury Department (Department of Alcohol, Tobacco, and Firearms)
- Explosives User Permit

Mine Safety and Health Administration
- Mine Identification Number (1)
- Legal Identity Report (1)
- Miner Training Plan Approval

TABLE C-4, Continued

STATE OF WASHINGTON

Washington Department of Ecology	• National Pollutant Discharge Elimination System (NPDES)/Construction Activities Stormwater General Permit • State Waste Discharge Permit • Water Quantity Standards Modification • Water Quality Certification (Section 401 - Federal Clean Water Act) • Dam Safety Permits • Reservoir Permit • Permit to Appropriate Public Waters • Changes to Existing Water Rights • Notice of Construction Approval (Air Quality) • Air Contaminant Source Operating Permit • Prevention of Significant Deterioration (PSD) - (Air Quality) (2) • Dangerous Waste Permit (2)
Washington Department of Natural Resources	• Surface Mine Reclamation Permit • Forest Practice Application • Burning Permit (Fire Protection)
Washington Department of Fish and Wildlife	• Hydraulic Project Approval
Washington Department of Community Development, Office of Archaeology and Historic Preservation	• Historic and Archaeological Review (Section 106 National Historic Preservation Act of 1966)

TABLE C-4, Continued

Washington Department of Health
- Sewage Disposal Permit
- Public Water Supply Approval

Washington Department of Labor and Industries
- Explosive License
- Safety Regulation Compliance (1)

LOCAL GOVERNMENT

Okanogan County Planning Department
- Shoreline Substantial Development Permit
- Conditional Use Permit/Zoning Requirements
- Building Permits
- Maximum Environmental Noise Levels (1)
- Socioeconomic Impact Analysis Approval (County Commissioners)
- Growth Management Critical Areas Regulations

Okanogan County Health District
- Solid Waste Handling
- Septic Tanks and Drain Field Approval

Okanogan County Public Works Department
- Road Construction and/or Realignment

Okanogan Public Utility District (PUD)
- Power Service Contract

SOURCE: Final EIS, Crown Jewel Mine, Washington (Forest Service/WSDE, 1997)

NOTES: (1) Performance standard/requirement - No formal permit necessary.
(2) Potential permit - At this time, these permits are not anticipated for the Crown Jewel Project.

TABLE C-5 Ruby Hill Project, Nevada, January 1997

Permit/Approval	Granting Agency
Approval of Plan of Operation	Bureau of Land Management
Surface Disturbance Permit (Air Quality)	Nevada Department of Conservation and Natural Resources, Division of Environmental Protection, Bureau of Air Quality
Permit to Operate (Air Quality)	Nevada Department of Conservation and Natural Resources, Division of Environmental Protection, Bureau of Air Quality
Water Pollution Control Permit	Nevada Department of Conservation and Natural Resources, Division of Environmental Protection, Bureau of Mining Regulation and Reclamation
Reclamation Permit	Nevada Department of Conservation and Natural Resources, Division of Environmental Protection, Bureau of Mining Regulation and Reclamation
Permit to Appropriate Water	Nevada Department of Conservation and Natural Resources, Division of Water Resources
Industrial Artificial Pond Permits	Nevada Department of Conservation and Natural Resources, Nevada Division of Wildlife
Approval to Operate Sanitary Landfill	Nevada Department of Conservation and Natural Resources, Division of Environmental Protection, Bureau of Solid Waste
General Discharge Permit (Stormwater)	Nevada Department of Conservation and Natural Resources, Division of Environmental Protection, Bureau of Water Pollution Control
Hazardous Materials Storage Permit	State of Nevada, Fire Marshal Division

SOURCE: Final EIS, Ruby Hill Project, Nevada (BLM, 1997a).

TABLE C-6 Lisbon Valley Copper, Utah, February 1997

Agency	Item/Permit	Description	Submittal Data	Likely Permit Specifications/Comments
FEDERAL				
U.S. Bureau of Land Management	POO, EIS	Environmental report including all aspects of operation, environmental and socioeconomic impacts, and mitigation.	Submittal data include air quality, areas of critical environmental concern, cultural resources, prime or unique farmlands, floodplains, Native American religious concerns, threatened or endangered species, solid and hazardous waste, water quality, wetlands and riparian zones, wild and scenic rivers, wilderness, paleontology, and other issues.	BLM as lead agency. Because of the location, and environmental sensitivity of the project, an EIS is required. A permit is not issued; approval of a selected alternative is granted in the form of Record of Decision (ROD). The BLM has a Memo of Understanding (MOU) with the Utah Division of Oil, Gas and Mining (UDOGM) concerning mine permitting and bonding.
	Right-of-Way	Right-of-Way grant required for authorization of power line.	PacifiCorp submitted Right-of-Way application specifying location and use.	Avoid cultural resource sites during construction.
U.S. Environmental Protection Agency	National Pollution Discharge Elimination System, (NPDES) - Water Quality	Must comply with surface and groundwater quality standards for discharge and non-discharging systems. State of Utah Department of Environmental Quality has EPA primacy for issuance of these permits.	Application fee and a characterization of baseline conditions, surface water, and groundwater hydrology.	To control discharge of metals and other potential effluents. Monitoring of discharge and reporting would be required.

TABLE C-6, Continued

Agency	Item/Permit	Description	Submittal Data	Likely Permit Specifications/ Comments
	Prevention of Significant Deterioration (PSD) - Air Quality	Permit is required if the operation of the proposed facility would emit greater that 250 tons of both point source and fugitive emissions from the facility. State of Utah Department of Environmental Quality has EPA primacy for issuance of these permits.	This environment evaluation includes all climatology and air quality data and identification and evaluation of all sources of fugitive and point source emissions, and a modeling of those emissions to project air quality impacts.	A permit is issued to control emissions of hazardous air pollutants, visible emissions, particulate emissions, and sulfur emissions. Monitoring and reporting is required.
U.S. Fish and Wildlife Service	Threatened and Endangered Species	Must prepare Biological Opinion based on projected impacts to threatened and endangered species in area of project. Comply with Section 7 consultation with ESA.	Preparation of Biological Assessment by BLM precedes Formal Section 7 consultation conducted as a part of EIS.	A permit is not issued; USFWS and State wildlife agencies use EIS as resource document to demonstrate compliance.
U.S. Mine Safety and Health Administration	Safety Permit	Must address operational safety issues.	Compliance with health and safety requirements	Identification number assigned.
U.S Army Corps of Engineers	Section 404 Permits - Dredge and Fill Activities in Watercourses	Provides protection for wetlands by regulating dredged or fill disturbances.	Submit water quality and other environmental data and development data.	Required for stream diversions and wetlands disturbance; compliance with Nationwide Permit 26.

TABLE C-6, Continued

Agency	Item/Permit	Description	Submittal Data	Likely Permit Specifications/Comments
STATE OF UTAH				
Department of Environmental Quality				
Division of Water Quality	Groundwater Discharge Permit	This permit is required for all activities having the potential to affect groundwater. Primacy action for EPA under provisions of CWA	A permit application is required that shows all water-discharging facilities and their design, along with proposed monitoring requirements.	Compliance with all Federal, State and local water quality parameters or site-specific standard based upon groundwater monitoring.
	Stormwater Approval Order	To satisfy stormwater permitting requirements on the state and federal land. Primacy action for EPA under provisions of CWA.	Construction and operation permit required for stormwater discharges. Application fees required.	BMPs (Best Management Practice) would be required.
Division of Air Quality	Air Quality Approval Order	Required for the construction of any facility or activity that may emit both a point source and a fugitive emission.	Submit permit dust control plan application that describes volume of through put and the location of proposed disturbance activities.	For compliance with Federal and State air quality point source requirements for both mining operation and powerline construction.
Division of Drinking Water	Public Water Supply Permit	Required for projects with more than 25 employees.	This permit requires design and control systems for clean drinking water, septic tanks, leach fields, and a review of any proposed landfill at the project area.	Includes regular monitoring of an on-site water supply or purchase orders if drinking water is provided from an outside source.
Division of Environmental Response and Remediation	Permits for Underground Storage Tanks	Permits required if underground storage tank or tanks are proposed.	Design specification of proposed tanks along with a description of the hydrology of the project area.	Independent monitoring and leak detection would be required.

TABLE C-6, Continued

Agency	Item/Permit	Description	Submittal Data	Likely Permit Specifications/ Comments
Division of Radiation Control	Radiation Control Permit	For the operation of equipment with radioactive material.	The specifications of the proposed equipment, the location of proposed equipment, and training and responsible party information.	Annual reporting and calibration reports.
Division of Solid and Hazard Waste	Resource Conservation and Recovery Act (RCRA) Permit	Permit to build and operate any type of solid waste disposal facility.	An analysis and characterization of all proposed waste products that would be disposed of (this may include waste dump material).	If there is hazard constituency to the proposed solid waste, there may be a requirement for lime facilities. There would be a requirement for annual reporting of volume placed in the facilities.
Department of Natural Resources				
Division of Oil, Gas, and Mining	Notice of Intent to Conduct Mining Operations Approval of Bonding	A proposed plan of mining operations, reclamation plan, and environmental impacts.	An application fee, environmental description, a mining plan, and reclamation plan.	Annual reporting requirements of production as well as reclamation activities and bonding requirements. An MOU is in place with the BLM to address bonding and other issues.
Division of State Lands and Forestry (2)	Lease	Must address all impacts on state lease lands.	Plan of Operations, reclamation plan, proposed bond to guarantee reclamation, and a schedule.	Annual fees and a report on throughput and reclamation activities.
Division of Water Rights	Water Right Permit	This permit requires an appropriation for a beneficial use, of which mining is considered to be a primary use.	A filing fee, well location, and information on surrounding appropriations.	Annual reporting requirements of volume of water used, and water level monitoring.

TABLE C-6, Continued

Agency	Item/Permit	Description	Submittal Data	Likely Permit Specifications/ Comments
	Impoundment Permits	Approval for any impoundment (dam) or the storage of water or solution.	Impoundment design specifications.	Leak detection monthly, quarterly, and annual reports as well as water level information.
Division of Wildlife Resources	Vegetation and Wildlife Impacts	Review of mining impacts on Federal and State listed sensitive species, as well as threatened and endangered species.	Information on surface disturbances, as well as a review of the reclamation plan to ensure compliance with surrounding vegetation and wildlife utilization; as a part of the EIS.	No formal permit required. Recommendations for mitigation may be incorporated in final BLM/DOGM approvals.
Other Agencies				
Utah State Historic Preservation Office	Compliance with the NHPA	A review of project area for significant archaeological and historic sites.	A cultural resources report showing the results of literature review, field surveys, and NRHP (National Register of Historic Places) significance evaluation.	Mitigation of any potential adverse effects to Federal and State significant sites.
Local Health Department	Septic Permit	Sanitation disposal permit.	Construction details with design capacities must be reviewed and approved by regional (state) health representative.	Facility must be sized adequately to successfully treat/handle sanitary waste.

SOURCE: Final EIS, Lisbon Valley Copper Project, Utah (BLM, 1997b).

NOTES: (1) Adapted from information provided by Summo USA Corporation. This list may not be all-inclusive; the operator is responsible for securing all the necessary permits and approvals for the project. (2) Mining activities that would occur on State lease lands. Division of Oil, Gas, and Mining has primary state lead on review of mine plan, reclamation plan, and bonding.

TABLE C-7 Olinghouse Mine, Regulatory Responsibilities

Authorizing Action/Permit/Regulatory Requirement	Regulatory Agency
Plan of Operations	U.S. Department of the Interior, Bureau of Land Management (BLM)
National Environmental Policy Act	BLM and Environmental Protection Agency (EPA)
National Historic Preservation Act	BLM and Nevada State Historic Preservation Office
Native American Graves Protection and Repatriation Act	BLM
American Indian Religious Freedom Act	BLM
Environmental Justice	BLM
Clean Water Act (Section 404)	U.S. Army Corps of Engineers (USCOE)
High Explosive License/Permit	Bureau of Alcohol, Tobacco, and Firearms
Industrial Artificial Pond Permit	Nevada Division of Wildlife (NDOW)
Water Appropriation Permits	Nevada State Engineer, Nevada Division of Water Resources
National Pollution Discharge Elimination System (NPDES) Permit	Nevada Division of Environmental Protection (NDEP)
401 Certification	NDEP
Surface Disturbance Permit (Air Quality)	Washoe County Health Department, Air Quality Division
Permit to Construct (Air Quality)	Washoe County Health Department, Air Quality Division

TABLE C-7 Continued

Authorizing Action/Permit/Regulatory Requirement	Regulatory Agency
Permit to Operate (Air Quality)	Washoe County Health Department, Air Quality Division
Water Pollution Control Permit	NDEP
Mine Reclamation Permit	NDEP
Solid Waste Disposal Permit	NDEP
Potable Water	NDEP
Sewer System Approvals	Nevada Department of Health, NDEP
Safety Plan	Mine Safety and Health Administration (MSHA)
Threatened and Endangered Species Act	U.S. Fish and Wildlife Service (USFWS)
Hazardous Materials Permit	Nevada State Fire Marshal
Special Use Permit	Washoe County
Building Permit(s)	Washoe County
Encroachment Permit	Nevada Department of Transportation (NDOT)

SOURCE: Final EIS, Olinghouse Mine, Nevada (BLM, 1998).

TABLE C-8 Yarnell Project, Arizona, June 1998: Regulatory Compliance Summary

Regulatory Agency	Law, Regulation, Permit, Document	Requirements
U.S. FEDERAL AGENCIES		
Bureau of Land Management	Federal Land Policy and Management Act	Approved operations conform to FLPMA requirements
	Mining Plan of Operations (MPO)	Approval by the BLM and reclamation bonding
	Use and Occupancy Regulations	Concurrence with regulations
	Reclamation Plan Requirements	Development, approval of reclamation plans and financial security
	Cyanide Management Plan 1992	Compliance with operational guidelines; required inspections
	NEPA Environmental Analysis	EIS prepared by the BLM as lead agency and Record of Decision on MPO
	Authorization of water supply facilities on federal land	Approval as part of the MPO
	Endangered Species Act	Biological assessment and consultation with U.S. Fish and Wildlife Service
	Migratory Bird Treaty Act of 1918	Protection of migratory birds
	Executive Order 11990	Protection of wetlands
	National Historic Preservation Act	Evaluation and consideration of project effects on properties eligible for the National Register of Historic Places, Native American consultations
	Executive Order 12898	Environmental justice in minority and lower income populations
	Executive Order 13007	Consideration of project effects on Indian sacred sites
	Department of Interior Secretarial Order 3175	Consideration of project effects on Indian Trust Resources

TABLE C-8, Continued

Regulatory Agency	Law, Regulation, Permit, Document	Requirements
U.S. Environmental Protection Agency (EPA)	National Pollutant Discharge Elimination System (NPDES) Permit (Clean Water Act)	Permit required to discharge to surface water from point sources other than process areas using cyanide
	NPDES Stormwater Discharge Permit (Clean Water Act)	Permit and Stormwater Pollution Prevention Plan required for monitoring and best management practices to reduce stormwater pollution discharge
	Section 404 Permit (Clean Water Act)	Consultation and oversight responsibilities with COE
	NEPA Environmental Analysis	Cooperating agency in preparation of EIS
	40 CFR 112 Spill Prevention Control and Countermeasures (SPCC)	SPCC Plan required for inspection of petroleum storage and dispensing facilities and actions to be taken in the event of a release of oil or fuel on-site
	Clean Air Act	Review and concurrence of state issued permit
U.S. Army Corps of Engineers (COE)	Section 404 Permit	Wetland and jurisdictional waters delineation, protection, and mitigation
U.S. Fish and Wildlife Service (USFWS)	Endangered Species Act	Threatened or endangered species evaluation
	Fish and Wildlife Coordination Act	Consultation with COE on Clean Water Act Section 404 Permit
Mine Safety and Health Administration (MSHA)	Health and safety regulations	Training and compliance during operations

TABLE C-8, Continued

Regulatory Agency	Law, Regulation, Permit, Document	Requirements
STATE OF ARIZONA AGENCIES		
Department of Environmental Quality (ADEQ)		
Air Quality Division	Air Installation Permit/Permit to Operate (Clean Air Act)	Permit related to construction and operational activities
Aquifer Protection Permit Unit	Clean Water Act	Section 401 Water Quality Certification
	Aquifer Protection Permit	Permit specifying process solution containment features and monitoring requirements for groundwater protection
		Contingency Plan required for actions to take in the event of a release of chemicals or process water from the site facilities
Department of Agriculture	Salvage or Removal Permit	Salvage or removal of protected native plants
State Historic Preservation Office (SHPO)	National Historic Preservation Act	Evaluation of project effects on cultural and historic resources
Mine Inspector's Office	Arizona Mining Code	Training and operations to conform to regulations
	Mined Land Reclamation Act	Reclamation/mine closure/bonding
Department of Transportation	Use Permit	Detailed traffic control plan to coordinate emergency services
Department of Public Safety	Notification required for state highway closure	Notification to State Patrol required to stop traffic on Highway 89 for blasting

TABLE C-8, Continued

Regulatory Agency	Law, Regulation, Permit, Document	Requirements
STATE OF ARIZONA AGENCIES		
Department of Environmental Quality (ADEQ)		
Air Quality Division	Air Installation Permit/Permit to Operate (Clean Air Act)	Permit related to construction and operational activities
Aquifer Protection Permit Unit	Clean Water Act	Section 401 Water Quality Certification
	Aquifer Protection Permit	Permit specifying process solution containment features and monitoring requirements for groundwater protection
		Contingency Plan required for actions to take in the event of a release of chemicals or process water from the site facilities
Department of Agriculture	Salvage or Removal Permit	Salvage or removal of protected native plants
State Historic Preservation Office (SHPO)	National Historic Preservation Act	Evaluation of project effects on cultural and historic resources
Mine Inspector's Office	Arizona Mining Code	Training and operations to conform to regulations
	Mined Land Reclamation Act	Reclamation/mine closure/bonding
Department of Transportation	Use Permit	Detailed traffic control plan to coordinate emergency services
Department of Public Safety	Notification required for state highway closure	Notification to State Patrol required to stop traffic on Highway 89 for blasting

TABLE C-9 Yarnell Mining Project, Arizona, June 1998: Sigificant Issues Raised During Scoping

Issue Category	Issues
Water Resources	• Impacts on the quality of surface waters in the watershed, both during the life of the mine and after the mine closes
	• Potential changes to the quantity of surface water flows as a result of groundwater pumping by the mine
	• Impacts on the quality of groundwater and water in wells in Glen Ilah, Yarnell and the surrounding area, both during the life of the mine and after the mine closes
	• Potential for depletion of the water table and wells as a result of groundwater pumping
	• Potential accumulation of water in the mine pit and the quality of that water during the life of the mine and after the mine closes
Air Quality	• Impacts resulting from dust, fumes, and chemical emissions
	• Potential for cyanide emission release
	• Public health issues associated with airborne transmission of disease, dust, or emissions
Blasting	• Impacts on the stability of natural features including boulders and aquifer systems
	• Potential for damage to residences, utility lines, and roads
Noise	• Impacts on public health and the quality of life in the nearby communities
Visual Resources	• Impacts on views from residences and Highway 89 during the life of the mine and after the mine closes
	• Effects of lighting on the night sky
Public Safety and Transportation	• Potential hazards created by truck traffic and the transport and storage of hazardous materials
	• Potential hazards to motorists from blasting
	• Effects of road closures on access to medical and emergency services by area residents

TABLE C-9, Continued

Issue Category	Issues
Socioeconomic Conditions	• Impacts on property values • Impacts on employment and income • Impacts on local businesses • Impacts on tourism • Impacts on tax revenues • Impacts on crime rates • Potential for increased demand on local services from possible influx of mine employees • Disruption of quality of life from noise, visual impacts, night lighting, or other aspects of the mine operation
Closure and Reclamation	• Adequacy of bonding to ensure completion of reclamation • Effectiveness of proposed reclamation plan and monitoring measures
Biological Resources	• Impacts to wildlife and wildlife habitata • Impacts to threatened or endangered species • Potential wildlife mortality from exposure to hazardous substances • Impacts on vegetation including riparian zones along Antelope Creek
Cultural Resources	• Impacts on prehistoric or historic sites and roads
Land Use	• Impacts on livestock grazing, other land uses and access routes

SOURCE: Draft EIS, Yarnell Mining Project, Arizona (BLM/EPA 1998).

Appendix D

Research Needs

The objective of the research discussed below is to minimize adverse environmental effects of mining on federal lands by filling gaps in knowledge about the long-term environmental impacts from hardrock mining. This appendix expands on the information provided in Chapter 3.

Research is needed in a number of areas and is summarized in this appendix according to scientific discipline. The areas of highest priority are listed below and may encompass more than one of these disciplines.

- **Pit Lakes:** A number of significant research needs are related to the long-term environmental impacts of pit lakes. Making accurate long-term projections based on short-term data is challenging, and the concordance between predicted and actual outcomes has not been evaluated. Research on the chemistry, hydrology, and biology of pit lakes and their surroundings is needed to minimize the environmental impact of those that presently exist and to improve the design of those proposed for the future.

- **Acid Drainage:** Improved methods for predicting and preventing acid drainage are also a high-priority research need. Acid drainage is the source of many of the water quality problems associated with hardrock mining. Improved methods for prediction, prevention, and long-term treatment are needed to minimize the expenses related to acid drainage and to enhance the long-term protection of the environment.

- **New Technologies:** Research on the mining approaches of the future is the next most important research area. Mining methods such as bioleaching and *in situ* mining, which are being proposed for a number of new mining operations, have the potential to prevent pollution, yet the long-term environmental consequences of these methods have not been investigated.

Each topic area requires three types of research:

(1) modeling to accurately project the changes in environmental parameters forward in time and outward in space;
(2) monitoring to provide information on the averages, extremes, and trends of environmental indexes, and to help calibrate and verify the accuracy of the models; and
(3) sampling and testing to acquire baseline data in and around mines.

It is impossible to predict which types of research will provide answers to the needs identified here or whether the critical problems have been identified. The scientific community needs support and flexibility to innovate and respond to new opportunities, as well as to resolve the well-identified problems.

WATER QUALITY

Water represents by far the most important interface between mining and the environment, and its quality, quantity, and distribution provide some of the most effective criteria for monitoring the state of the environment. There is an array of models and laboratory techniques used to predict water quality and quantity from the leaching of mine waste and rock at mine sites. These techniques and models include tests for acid-generation potential, leach tests, pit lake models with water quality and quantity components, and waste rock and tailings discharge models with water quality and quantity components. The results from these tests and models are used to determine if pit water or leachate from mine waste will pose a threat to aquatic life, wildlife, or human health. Predictions of long-term water quality related to acid drainage, especially in pit lakes, have a high degree of uncertainty. Without reliable forecasts of long-term water quality, it is difficult to design effective mine waste management techniques to protect against future deterioration of water quality. Uncertainty about long-term water quality and quantity predictions points to a number of research needs that could help increase the accuracy or define the appropriate use of these predictive tools.

Pit Lake Water Quality

The objective of pit lake models is to accurately predict the chemistry and hydrology of pit water as the lake forms, after water levels have stabilized and taking into account any long-term impacts of evaporation or other factors that may affect water quantity. The modeling results can indicate the factors that control water quality and quantity and opportunities to adjust operation and closure plans so that water quality problem can be reduced. Pit lake models predict future water chemistry based on such factors as up-gradient groundwater concentrations, the amount and types of materials (including mineral phases), availability to weathering, leaching of materials, hydrological parameters, and oxidation and reduction reactions that may occur in the lake.

It is currently not known how accurately pit lake models predict concentrations of contaminants in pit lakes and surrounding groundwaters. Thus, comprehensive comparisons of predicted and actual concentrations in pit lakes and groundwaters are needed to evaluate whether existing models can predict long-term water quality. For example, a number of pit lakes at uranium mines have filled, and although pit lake models were rarely used before the lakes filled, it would be possible to model retrospectively the pit water chemistry and compare modeled results to actual concentrations. The parts of the models that needed improving would be adjusted based on results from the comparison study. Pit lakes in Nevada that are currently filling should be monitored to evaluate how the chemistry changes with time.

The use of a pit lake model without data from long-term leach tests increases the uncertainty of the predictions. Eary (in press) found that, for the toxic metals copper, cadmium, lead, and zinc, concentrations were not well represented by theoretical solubilities of known mineral phases, indicating the importance of empirical data based on adequate leach tests for predicting pit water quality. The Nevada Bureau of Mines has conducted an inventory of pit lake water quality in Nevada (Price et.al., 1995). This type of information can be used, in conjunction with knowledge of the geology and mineralization of the mines, to compare predicted and actual concentrations in pit lakes. Tempel et al. (in press) found that predicted arsenic concentrations in the North pit at the Getchell Mine in Nevada did not match actual concentrations. The lower measured concentrations in the pit water may be related to adsorption to aluminum hydroxide and clay mineral surfaces. Until more of these comparisons are conducted, the validity of pit water prediction models will remain uncertain.

The possibility that backfilling of pits will become more common raises additional research questions, because the environmental consequences of backfilling have not been adequately investigated. Partial backfilling of pits with acid-generating wastes has led to temporary increases in metal concentrations and acidity of pit waters, and geochemical models used to predict pit lake chemistry have had to take these changes into account (e.g., at the Sleeper Mine in Nevada) (Mark Logsdon, personal communication, 1999). The impact of partial or complete backfilling on down-gradient groundwater should also be investigated.

Acid Drainage and Leaching of Mine Materials

The prediction of both acid drainage and the leaching of mine waste and pit wall rock needs improving. Like pit lake models, there has been little effort to compare predicted and actual concentrations. Predicted concentrations of acid, metals and other constituents in waste rock, heap, and tailings discharge should be compared to measured concentrations to determine and improve the accuracy of these methods.

Sample collection strategies for estimating both acid drainage and mine waste leachate need re-evaluation. Sulfide ore bodies, which generally have the greatest potential to generate acid, are known to be heterogeneous in their metal and sulfide content. Therefore, proper sampling of all geologic materials that will become exposed to the environment or become waste is essential. The recommended number of samples per rock type or waste unit varies widely, from one sample for every million tons to one sample per 20,000 tons of waste rock or 50 samples for each million tons (Schafer, 1993; Forest Service, 1992). Another approach is to allow the variability of the material itself to dictate how many samples are collected and analyzed, and then to quantify and evaluate the results using statistical methods (Runnells et al., 1997). Compositing of samples can underestimate the variability in acid-generation potential; therefore, it is important to composite materials only of similar lithology and mineralogy. Waste rock is much more heterogeneous than tailings, and compositing of samples may be more acceptable. In addition to sampling of solids, long-term monitoring strategies for leachate need to be improved. A better understanding is needed of when and how to sample effluents, especially under variable seasonal and meteorological conditions.

Acid Drainage

Acid drainage forms when oxygen and water come in contact with sulfide minerals and certain metal sulfates that form from the weathering of metal sulfides. Factors affecting acid generation include sulfide amounts and types, particle size of waste material, pH, oxygen availability and diffusion, temperature, storage (evaporative concentration of acid products), availability and type of neutralizing material, water saturation, amount of ferric iron present in the water, and the presence of iron and sulfur-oxidizing bacteria (EPA, 1994). Although many states require the use of acid-generation prediction testing as part of the permitting process, variability in one or more of the factors listed above over time and the type of testing performed can limit the reliability of the test and projection results.

Although acid drainage at ancient mine sites has continued for millennia, the study of acid generation and associated drainage is recent. In addition, the generation process is not well enough understood to determine definitively how long it will run and how high the concentration of acid and metals will reach. Accurately predicting acid drainage potential is a very important element of accurately projecting pit lake water quality. Because of the importance of acid drainage, all aspects of the acid drainage assessment process could benefit from additional research.

Two general types of tests are used to predict acid-generation potential: static and kinetic. Static tests estimate the maximum acid-generation potential (AGP) and neutralization potential (NP) of a rock or waste material. Kinetic tests are conducted for six weeks or longer, use larger sample volumes, provide information on the acid production rate and drainage water quality, and are generally more reliable than static tests. Modified humidity cell- and column-type tests are currently the preferred kinetic tests (EPA, 1994). Special equipment is required, and the costs of the tests are higher than those of static tests. The opportunity to examine changes in pH and metal concentrations over time is available with the kinetic tests.

There are a number of instances where acid drainage has occurred even though it was not predicted or expected. Examples include the Thompson Creek mine in Idaho, the LTV Steel Mining Company in Minnesota (EPA, 1994), the Newmont Rain facility in Nevada (EPA, 1994), and the Zortmann-Landusky mine in Montana (Federal Register, August 7, 1996, [61 FR 41182]). In some cases, waste rock or pit wall material was not initially acid generating, but waste material removed later was. In other cases, static or kinetic tests results showed that acid generation would not be a problem, but actual drainage from mine waste units was acidic. In still other cases, the acid-base accounting data showed the potential for acidification, but errors in construction of the piles or errors in interpreting and reporting the data led to on-the-ground problems.

The effectiveness of the acid-generation prevention and source treatment methods is unknown. General research is needed into the effectiveness of the various treatments on various types of ores and waste. Research should focus on better understanding the acid-generation process and on the effectiveness of the prevention methods in the short and long term. Research is also necessary on comparison of modeled results, either by computer or in the laboratory, to actual field conditions at sites where the methods are in use.

Leaching of Mine Materials

The leach tests most commonly used were developed primarily to predict contaminant concentration in leachate from solid and hazardous waste landfill. Substantially less effort has been devoted to determining their accuracy in mining situations.

Leach tests used currently include the WET test, the toxicity characteristic leaching procedure (TCLP), the synthetic precipitation leaching procedure (SLP), and humidity cell tests (EPA, 1978). The WET, TCLP, and SLP tests are all short-term tests; the humidity cell tests can last for six weeks or longer. The leaching reagent (acetic acid) used in the TCLP may be appropriate for municipal landfills, but is not appropriate for mining wastes. The SLP test uses a lower pH leach solution for waste materials east of the Mississippi River, because of acid rain from the use of high sulfur coals.

The WET, TCLP, and SLP tests all may underestimate leachate concentrations because they are conducted in time frames that do not allow equilibration of materials with the leachate solution. Alternating wetting and drying used in some humidity cell tests best approximates the conditions that are most favorable for the formation of acid drainage. In some cases, it may take months or years for stable pH and metal concentrations to be reached. If the tests are cut short before steady state concentrations are reached, predicted concentrations may underestimate actual concentrations.

Finally, the fate and transport of leached metals and other constituents in the environment need additional research. For example, although cyanide and some cyanide-metal complexes are susceptible to photolytic degradation in surface water, not much is known about how they behave in down-gradient groundwater that eventually may be discharged to surface water or be used for drinking water or irrigation.

HYDROLOGY

Like long-term water quality predictions, the modeling of water quantity and hydrologic processes also contains uncertainties. It may not be known, for example, whether some pit lakes will have closed-basin or flow-through hydrologic features, or whether and under what circumstances they may turn over. Long-term ecological, water quantity, and water quality impacts of pit dewatering and discharge of pit water to streams are beginning to be investigated, but results of these studies are not yet available. Site-specific water balances, which in part were responsible for uncontrolled discharges from the mine at Summitville, Colorado, are not sufficiently understood. The flow of water through and interaction with unsaturated waste rock and saturated fractured media is poorly characterized and affects the validity of models that predict the behavior of groundwater in these media. Water quantity affects water quality, and uncertainties in one area compound uncertainties in the other. The following research areas in hydrology warrant further investigation.

Pit Lake Water Quantity

The most important physical parameters for a pit lake model are evaporation, precipitation, groundwater flow, and transfer of gases into and in the lake. These processes affect lake level, whether or not the lake turns over, and the oxidation and reduction state of the pit water, all of which affect lake chemistry. There are many methods that can be used to estimate evaporation rates for the pit lake. Pan evaporation rates may overestimate or underestimate evaporation depending on the use of correction factors, the wind speed across the lake, and other factors (Bird, 1993). Wind speed, the amount of precipitation and surface runoff entering the lake, and the amount of metals and other constituents associated with run-on water will change from year to year depending on the amount of precipitation and other climatic factors. Long-term weather data may not be available to provide accurate maximum and minimum precipitation amounts. The original groundwater flow pattern around a pit may be altered by dewatering or by the presence of the lake, which presents large amounts of water for evaporation that were previously present as groundwater. Large precipitation events could cause the lake level to rise above that of the surrounding groundwater and discharge potentially contaminated pit lake water to groundwater (Macdonald et al., 1994). The long-term hydrologic status of pit lakes is affected by all the physical factors discussed above. Whether a lake is a flow-through system or a closed-basin lake with no outputs will affect the chemistry of the lake water. If evaporation is significantly greater than precipitation, the concentrations of metals and other constituents will increase over time through evapoconcentration. If, however, the lake flows through for all or part of the year, down-gradient groundwater concentrations may increase above pre-mining concentrations.

Mine Area Dewatering and Discharge of Surplus Water

Mining influences surface flows on and off site through alteration of runoff from several types of constructed surfaces (e.g., roads, waste piles, leach pads), through discharge of surplus water from pit or underground dewatering, and through use of water for processing. Groundwater withdrawals for processing and pit dewatering may affect local and regional aquifers. Pits can influence local hydrology through interaction with the local aquifers, interception of runoff and precipitation, and lake surface evaporation. Mines may supplement stream flow with surplus water from pit dewatering, and then, after mining operations cease, diminish stream flow as the pits and dewatered aquifers are allowed to fill.

The impacts of mine area dewatering include reduced flow in springs, which in arid areas provide habitat for terrestrial wildlife. The flow of streams may also be decreased, which can change the type of aquatic biota that can survive in that system. A decrease in streamflow can increase concentrations of contaminants in the stream, especially if mine pollution sources discharge directly or indirectly to surface water. The existence of a pit creates a permanent higher evaporation sink and can lower the water table off site, especially in arid areas such as Nevada.

Modeling of groundwater withdrawal (Schaefer and Harrill 1995) has demonstrated the effects of pumping from deep and shallow aquifers on shallow basin-fill water tables. Other studies have shown reduction in local stream flow when water is withdrawn from nearby deep aquifers. Water balance models for different hydrogeological settings should be developed that address local and regional interrelationships among surface flow, pit lake hydrology, and hydraulic head of shallow and deep aquifers. These models should enable long-term predictions of the consequences of alteration of surface waters, and interruption, use, and withdrawal of groundwater by mining activities. The models should include consideration of transient events.

Surface water discharge may alter the timing of high flow events, which may eliminate recruitment of riparian vegetation and eventually lead to senescence and death of existing vegetation. This can increase erosion of streambanks, which provide shelter and shade for fish. Long-term hydrologic, geomorphic, and water quality changes from riverine discharge of mine water should be investigated and modeled. Studies on riverine discharge of mine water to the Humboldt River in Nevada are under way at such federal agencies as the U.S. Geological Survey and the Fish and Wildlife Service, but similar studies elsewhere, and site-specific modeling, will facilitate long-term predictions of the response of riverine ecosystems to hydrological changes.

Modeling Flow in Mine Dumps and Impoundments

Waste rock dumps and tailings impoundments show strongly heterogeneous patterns of permeability, which current hydrologic models are incapable of reproducing. In the case of waste rock, end dumping from haul trucks produces inclined layering in the waste rock dump, with a pronounced gradation from the finest material at the top of the dump to very coarse material at the bottom. Compaction of the upper surfaces of the dumps by movement of haul trucks and other vehicles tends also to produce horizontal layers that may be less permeable than other portions of the dump and may act as barriers to fluid flow.

In the case of tailings impoundments, discharge of the tailings from spigots produces horizontal and vertical sedimentation of particles of different sizes. The coarsest materials are deposited near the points of discharge and the finest grained materials (slimes) are transported along a gentle slope down gradient from the points of discharge toward the center of the pond. As a result of this differential sedimentation, the horizontal permeabilities in a tailings impoundment tend to be much higher than the vertical permeabilities. These heterogeneities are not included in current hydrologic models and should be considered as improvements are made to these models.

The flow of water through unsaturated waste rock and saturated fractured media is also not well understood and affects the validity of models that predict groundwater flow. Improved models are needed to more accurately predict the infiltration and flow of water into waste rock dumps and tailings impoundments from rainfall and snow melt.

ECOLOGY

Modern hardrock mining is creating potential biological habitats in pit lakes. The viability of these lakes as long-term habitat and food sources for aquatic biota and wildlife has not been evaluated. In addition, the long-term sublethal effects of cyanide and metals on aquatic biota and migratory birds have not been extensively studied. The following areas need further study:

- potential development of biological communities in pit lakes and impacts on aquatic biota and wildlife;
- sublethal effects of cyanide and metals on aquatic biota and migratory birds;
- establishment of water quality standards for pit lakes and waste leachate for protection of aquatic biota and wildlife;
- integrated ecologic-hydrologic studies and modeling to determine the consequences of minewater discharge on riverine ecosystems; and
- impacts of placer mining on stream ecology and aquatic biota.

The viability of creating fish habitat in pit lakes turn in part on the long-term water quality predicted for the lakes. The hydrologic and water quality condition of the lake may change if it later becomes a closed-basin lake or begins to turn over. The nature of the food chain in the lake, such as the type of algae and bacteria, and its ability to bioconcentrate or accumulate contaminants has not been studied. Certain birds may feed on zooplankton in the pit lake, while others may feed on fish. Food chain effects in general in pit lakes need to be studied to determine the viability of creating long-term biological habitats in these mine features.

Sublethal (e.g., growth, reproduction, behavioral avoidance) effects of metals on salmonids and macroinvertebrates have been studied to some extent (see Appendix B, "Aquatic Biota"), but sublethal effects on wildlife, including migratory birds, have received less attention. The effects of cyanide and metal-cyanide complexes on aquatic life and wildlife have received less attention. If pit lake cyanide and metal concentrations meet chronic aquatic life criteria, aquatic biota will not die off immediately, but there may be a long-term effect for communities that live in the lake or feed on biota from the lake. Investigation into acceptable whole-body metal and cyanide burdens for consumption by migratory birds should also be investigated.

Water quality standards for pit lakes and mine waste leachate have not been designed for protection of migratory birds. Research is needed to establish the scientific bases for standards for consumption of lake water by waterfowl and migratory birds. In addition, leachate standards have only been developed for a limited number of constituents, and are based only on human health concerns. Therefore, research is needed to establish the scientific bases of leachate standards for protection of aquatic life in cases where leachate from a mine waste deposit has the potential to discharge directly or indirectly (through groundwater) to surface water.

The discharge of mine water to streams can increase loadings of contaminants, even if the concentrations are lower than those in the stream. The added influx of contaminants can be retained on stream sediments and periphyton, which serve as habitat and food sources for macroinvertebrates, which in turn are one of the main food sources for fish. In terminal basins, such as the Humboldt Sink in Nevada, discharge of mine water to streams that feed the sink will cause metals and other constituents to evapoconcentrate in the sink, which provides habitat for waterfowl and migratory birds.

Finally, the impact of placer mining on stream ecology should be investigated. Placer (mechanized) or suction dredge mining disturbs streambed sediments, which provide habitat for macroinvertebrates and spawning habitat for salmonids. The long-term impact of placer mining is unknown, especially under low-flow conditions or when springtime high flows are not able to return bed sediment to its original characteristics. As discussed in Appendix B, studies on the impact of suction dredge and mechanized placer mining on macroinvertebrate and spawning habitat should be conducted.

CUMULATIVE IMPACTS

A mine's environmental impacts are considered during the preparation of the environmental impact statement (EIS), but multiple, diverse demands on the resources of a region (e.g., a complex ecosystem or a watershed) can create interactions and interferences that may be beyond the purview of individual impact statements. If a proposed mine is near other mines or other human activities, there is a potential for cumulative impacts that extend beyond what an isolated mine may cause. Cumulative impacts are considered during the EIS process, but methodologies for accurately predicting and assessing these impacts are not well developed. Research is needed to improve our understanding of how mining participates in cumulative effects and to predict impacts from mining under different environmental circumstances.

An example of cumulative impacts includes the impact of exploration combined with hunting, off-road vehicle use, fishing, and camping that may use access roads created for exploration. Also, groundwater withdrawal by several mines may deplete a regional aquifer that is shared by agriculture or urbanization. Each incremental withdrawal introduces potentially expanded impacts that need to be evaluated during the EIS process. In addition, the cumulative tapping of groundwater that feeds springs may affect the wildlife and riparian habitat.

Research is needed on the cumulative impacts of past and current mining activity and the development of predictive models for use in evaluating proposed actions. Long-term monitoring should be designed to measure attributes that offer useful data on changes and sustainability of resources affected by mining activity and post-closure conditions. With respect to cumulative impacts, it is particularly important to develop long-term monitoring strategies for water quality and biological analyses that correspond with critical periods in the hydrograph.

ALTERNATIVE MINING AND POLLUTION PREVENTION METHODS

The potential environmental consequences, of new mining techniques should be assessed before they come into widespread practice. Some of these techniques have pollution prevention potential, but the net environmental impact of the practices discussed below generally has not been investigated. These techniques include bioleaching, deep underground mining, *in situ* mining, and the use of alternative lixiviants.

Alternative Mining Techniques

In the precious metals industry, especially in Nevada, the near-surface oxidized ore is being exhausted, and the industry is moving into the deeper sulfide reserves. Because of their sulfide characteristics, these reserves are not as amenable to cyanide recovery as the oxide ores. This problem is being addressed mainly by finding ways of economically oxidizing the ore so that cyanide leaching can be used, and by identifying substances (lixiviants) that can effectively leach the precious metal from the sulfidic ores.

Bioleaching is one way of oxidizing the ore. In this approach, the ore is treated with a low-pH solution into which iron- and/or sulfur-oxidizing bacteria are introduced. Typically, *Thiobacillus ferrooxidans* or *thiooxidans* (the same microbes responsible for acceleration of the acid drainage production rate) are issued and, not surprisingly, one of the by-products is sulfuric acid. After microbial oxidation, the ore is neutralized and put through the cyanidation process, either in a heap leach or a milling circuit.

Although bioleaching is similar to leaching of copper ores with sulfuric acid or the production of acid drainage, the combined environmental impacts of acidic drainage and cyanide have not been examined. For example, the kinetics of formation and the stability of metal-cyanide complexes may be different under acidic, rather than basic, pH values. In addition, the long-term acid generation potential of neutralized and cyanidized sulfidic ore has not been investigated. The long-term environmental impacts of bioleaching and other new techniques for recovering precious metals from sulfide ore bodies need to be investigated.

Much of the new mining in the United States may be deep underground mining (see Appendix A). Although there will be less surface disturbance, impacts on the quality and quantity of deep groundwater and other environmental factors have not been investigated. As with pit mining, these environmental impacts may be complicated by an increased emphasis on backfilling of the underground shafts and tunnels, which can affect the stability of the underground workings and drainage and groundwater chemistry.

In situ mining has been conducted for decades in the uranium industry, but the techniques have only recently been applied to base metal mining, especially copper mining. This process involves injecting leaching agents directly into the ground and collecting the pregnant solutions in down-gradient wells. Permitting these types of facilities has been difficult, in part because of the uncertainties associated with groundwater quality and quantity.

The use of alternative lixiviants has been investigated for years, but the long-term impacts of the alternatives on aquatic life, wildlife, and groundwater and surface water quality have not been sufficiently researched. Particular attention should be given to identifying alternative lixiviants that are more environmentally acceptable than cyanide to leach precious metals; they would be valuable assets to industry and comforting changes to the public.

Pollution Prevention and Treatment

Some of the new technologies have the potential to prevent the amount of pollution generated by mining activities. However, additional research is needed to discover new and improved existing methods for cost-effective long-term treatment of acid drainage, other types of mine leachate, and other types of water quality problems. Some of the techniques being developed to clean up hazardous waste sites may also be applicable to mining. Examples include the use of iron curtains to treat groundwater contaminated with acid drainage (Shokes and Moller, 1999), bioremediation techniques, and the use of constructed wetlands to remove contamination from waste waters. Both the active and passive treatment systems require long-term maintenance and can be expensive to operate.

A common method for treating in place to prevent acid drainage is to mix the acid-generating materials with a neutralizing material such as lime. The materials can be layered with lime or physically mixed either before or after disposal. The long-term effectiveness of this type of active treatment merits additional examination. Both active and passive treatment systems require long-term care (in some cases in perpetuity) and maintenance, and can be expensive to operate. The systems are difficult to standardize when each is treating a unique chemical mix of metals and salts. More research is necessary on treatment methods to determine whether there are better treatment options for contaminated water and to find improvements that can be made to help them function with less maintenance over the long term. Operators and regulators alike recognize that the most cost-effective control may be to prevent the drainage from occurring in the first place, because once drainage begins, there is very little that can be done except long-term treatment of the effluent. Environmentally acceptable alternatives, such as pre-treating incoming waters or controlling the activity of the acid-promoting biota, are worthwhile research goals.

MINING TECHNOLOGIES FOR THE FUTURE

Technologies of the modern mining industry are very different from those of the past. Mines now employ highly efficient, advanced production systems, coupled with innovative engineering designs to prevent or significantly mitigate environmental disturbance. Even so, the remarkable advances in mining-related technologies, the better understanding of our environment, and the application of innovative engineering designs have not eliminated all detrimental impacts of mining, nor are they likely to do so. Mining technologies of the future will continue to improve operational efficiency, productivity, and safety while minimizing pollution and maximizing control through improved designs and efficient operation. It is in these areas where additional incentives exist to develop innovative, if not revolutionary, minerals extraction technologies.

Future mine production systems will consist of people, machines, computers, sensors, and the communication links among them. New technologies will be developed that offer the potential for major discontinuous change in the way minerals and metals are discovered, mined, and processed. To extend the lives of U.S. mines and discover new mineral deposits, more and better technologies will be required as mineralized grades continue to decline, metallurgical characteristics of the ores become more complex, and environmental regulatory pressures build. Examples of new technologies are presented below (Beebe, 1995; Gentry, 1998).

Exploration

Advances in control mechanisms, mostly implemented with computers, and sensor development will enable minerals exploration to be far less intrusive to surface and subsurface environments, and include:

- satellite remote sensing techniques capable of penetrating high-density foliage, surface cover, and even short distances into the rock itself;
- advanced geophysics and geochemical techniques capable of detecting smaller and more subtle signatures of deeply buried deposits;
- borehole geophysics and other techniques to improve sampling, analysis, estimation of deposit reserves, and geotechnical assessments with a minimum of drilling into the subsurface; and
- manipulation of increasingly complex data bases for target and deposit modeling with sophisticated pattern recognition software that is both efficient and effective.

Mining

Although surface mining is the most efficient method for mining minerals in use today, its economies of scale and labor productivity may be approaching limits. With respect to surface area disturbed, surface mines are less benign than their underground counterparts; however, underground mining is generally more expensive in terms of capital and operating costs, more labor intensive, and more dangerous. Also, some underground mines can result in land and hydrologic disturbances due to surface subsidence and groundwater withdrawals.

In the future, there will be even greater pressures to increase mining efficiencies in terms of labor costs, productivity, capital investment, safety, and environmental sensitivity. Where these objectives cannot be achieved satisfactorily, social and environmental factors will tip the balance.

New mining technologies may provide the key to achieving the above objectives (Beebe, 1995; Gentry, 1998). There will be greater use of:

- remote-controlled and autonomous robotic equipment, progressing toward "intelligent" mining systems that require minimal human intervention;
- geosensing techniques to navigate and control intelligent mining systems to obtain assay and geotechnical data in real time and to warn of potentially dangerous variations, such as faults, shear zones, or pockets of water or gas;
- improved mechanical and some form of nonexplosive rock fragmentation techniques fitted to intelligent mining machines;
- *in situ* mining of small or deeply buried deposits economically and in an environmentally safe manner; and
- more efficient, effective, and safe underground extraction systems that comply with growing environmental pressures and meet society's demand for optimal life-cycle costing.

Mineral Processing

The tailings impoundments generated by mineral processing plants can preclude certain post-mining land uses or can release toxic constituents over time. The issue is to minimize the amount of tailings produced, find a suitable storage location for them, and subsequently stabilize and revegetate to the maximum extent possible.

It is frequently assumed that the relatively new advances in hydrometallurgical systems result in better waste minimization characteristics than the more traditional physical separations. This is true only if leaching agents can be entirely recycled or if discharge streams can be treated effectively to eliminate pollutants before discharge. Toxic leaching agents must be contained if subsequent environmental problems are to be avoided.

Future mineral processing and extractive metallurgy technologies must improve efficiencies in energy consumption, liberation and process metallurgy (including mathematical modeling for process optimization augmented by better sensors and control mechanisms), and most importantly waste reduction and stable disposal techniques. From an environmental standpoint, future technological advances in process metallurgy must address issues pertaining to methods that:

- generate less waste or create waste forms that are more stable or benign;
- establish more efficient separations that avoid discarding valuable or troublesome constituents;
- avoid dissolution of particularly troublesome metals or fix them as inert compounds;
- promote selective solution or dissolution of minerals through pressure oxidation and bioleaching technologies;
- promote more efficient, selective, and timely oxidation, and stabilization of minerals using bio-oxidation techniques and associated bioremediation techniques for environmental mitigation;
- use genetically engineered microbial approaches to oxidation, dissolution, and immobilization of metals and associated toxic elements;
- use hybrid processing flowsheets combining physical, chemical, and microbiological methods in place of repetitive stages of the same methods;
- use alternative lixiviants for *in situ* mining and processing technologies that are efficient in mineral recovery and are environmentally benign (i.e., agents that dissolve desired metals while minimizing dissolution of impurities);
- efficiently eliminate cyanide and other troublesome processing chemicals safely and rapidly;
- improve recovery of by-products or co-products, even ones considered unconventional by today's standards; and

- use waste treatments that minimize later dependence on time or natural processes to fix or eliminate troublesome constituents (Beebe, 1995; Gentry, 1998).

These technological advances should eliminate, or at least significantly mitigate, many of the environmental concerns now associated with mining and related technologies; however, it is doubtful that these future technologies will provide utopian solutions to environmental concerns. For example, *in situ* mining will present enormous challenges relative to lixiviant selection; solution and dissolution of desired versus undesirable metals; lixiviant containment and control in the structurally complex geologic settings hosting most hardrock mineral deposits; and related impacts on the neighboring groundwater regime.

Another example is the anticipated trend toward underground mining using automated, robotic mining systems. Such systems are likely to incorporate storage of significant amounts of waste rock and tailings underground in mined-out workings, but they will not eliminate surface disturbances or concerns about impacts to groundwater systems pertaining to waste isolation, containment, leakage, and impacts on hydrologic regimes. Surface subsidence will remain and will require equally sophisticated technologic systems to monitor and control.

Mining and associated exploration and processing technologies will advance, some quite rapidly (e.g., hydrometallurgy and biotechnology). While these new technologies will resolve many environmental concerns, they no doubt will introduce other areas of concern not yet recognized or contemplated. These technologic advances will present new opportunities and challenges to all stakeholders in the domestic minerals industry.

SUMMARY

In conclusion, research related to the long-term environmental impacts of mining is needed in the areas of water chemistry, hydrology, ecology, and mine treatment and pollution prevention technology. A number of important research areas have been discussed in this appendix. Results from a well-coordinated and consistently funded research program could help improve current methods used to predict, prevent, and minimize the environmental impacts of mining and could also reduce costs associated with long-term maintenance at mine sites.

Appendix E

Financial Assurance

Financial assurance is an important part of the regulatory framework to protect federal lands and the environment when an operator is unable or unwilling to perform reclamation and other obligations, and to address contingencies and future long-term post-closure management expenses. Financial assurance for mining operations raises four issues: (1) the activities and events the assurance covers; (2) the amount of the assurance; (3) the forms the assurance may take; and (4) conditions under which the financial assurance is to be released.

ACTIVITIES AND EVENTS COVERED

Financial assurance for reclamation of mining sites is part of mining regulatory requirements in all of the western states that have mining on federal lands. In addition, financial assurance for reclamation is required by the federal land management agencies for mining operations with approved plans of operations on federal lands. Notice-level operations under BLM regulations do not currently have to provide a financial assurance; under the regulations of some states, these small operations are also not required to post a financial assurance for reclamation.

Financial assurances required by plans of operations and state reclamation laws (and some state laws governing closure of waste management units) are monetary guarantees that provide for completion of the reclamation plan for a mining site when an operator is unable or unwilling to do it. Typically, these financial assurances also include money for monitoring the success of the reclamation program, including vegetation surveys and water quality monitoring. Several states have requirements for long-term maintenance of reclaimed units.

Several states—but not the Forest Service or BLM—have adopted regulatory programs that require or authorize the agencies to require financial assurance for long-term protection of water quality. Some mines on federal lands have provided such assurances negotiated on a case-by-case basis with state and federal agencies, even where explicit authority to require such assurance may not have been in the regulations.

AMOUNT OF FINANCIAL ASSURANCE

State laws, regulations, and state and federal guidance documents cover the determination of financial assurance amounts. Financial assurances for reclamation are calculated based not on the operator's cost to complete the work but on the cost to bring in a third party to complete the work with regulator agency and/or consultant oversight. The regulatory agency determines the costs, often based on information submitted by the operator. The equipment costs are determined based on equipment available for rental in the vicinity of the operation, fuel costs, and operator costs, typically an hourly rate. The equipment's efficiency at material movement and rate of progression of the work is determined from such resources as the Caterpillar Performance Handbook. Building demolition and disposal is based on standard demolition costs. Seed and revegetation costs are based on actual information available from seed distributors. Knowledge of local labor rates and typical time requirements to complete the work are also used to develop detailed costs for all aspects of the site reclamation.

Most financial assurance language allows for periodic review of the reclamation costs and the bond instruments to ensure that the financial assurance is adequate. Some regulations require that the full amount of financial assurance for the maximum disturbance be submitted prior to construction; others allow the financial assurance to be submitted for the amount to be disturbed at any one time, usually a year, plus the amounts previously disturbed and not successfully reclaimed prior to that time. Almost all financial assurances are for the life of the project, although regulatory language may allow for periodic review and for change to the amount or form of the assurance. For instance, financial assurance forms used in Colorado specify that the bond is for one year, but the bond is automatically renewable for subsequent years if no other bond is put in place prior to the expiration date. That allows the state to review the financial assurance on a yearly basis, but does not require them to do so in order to maintain the appropriate coverage.

TYPES OF FINANCIAL ASSURANCE

Regulations for each state and for the federal land management agency dictate types of financial assurances that are acceptable. Typically, cash, certificates of deposit, letters of credit, and corporate sureties are accepted in all settings. Less common, but still accepted in some states, are deeds of trust for real estate or liens on equipment. Some states also accept a guarantee known as self-bonding or corporate bonding for all or a portion of the financial assurance. This is a guarantee by the company that it will complete the reclamation work, and there is no exchange of money to the regulatory agency. The company provides information to show that it is a going concern and will be solvent for the life of the operation. Publicly or privately held trust funds established by companies have been used to guarantee long-term water treatment and monitoring.

Most states have regulations that dictate which state agency can hold the bond for reclamation. These regulations restrict the ability of other agencies in the state to hold a financial assurance for the same purposes. Where federal, private, or state lands are involved in an operation, the states and federal land management agencies have memoranda of understanding (MOUs) specifying which agency will hold the financial assurance. In most cases, the MOUs also determine how the assurance amount will be calculated and how it will be released.

CONDITIONS FOR RELEASE

Financial assurances for reclamation are required until reclamation is complete and successful. Reclamation is typically defined as the successful physical stabilization and revegetation of the site. Most states and the federal land management agencies consider partial release of the reclamation bond as portions of the work are completed. For example, if the operator has successfully completed regrading, that portion of the bond designated for regrading can be released.

Procedures for release of financial assurances vary among states. In most cases, the process is a technical determination by agencies responsible for administering the provisions guaranteed by the assurance. Some states, such as Colorado and Montana, expressly provide for receipt of comments on decisions to release reclamation bonds.

Trust funds for long-term monitoring and maintenance, such as for long-term water treatment, can be structured so that they are not released, but are administered by the public or private trustee to provide sufficient revenue to assure continuous funding.

Appendix F

Presentations to the Committee

Speakers at Committee Meetings

Stephen Alfers, Vice-Chairman for Hardrock Mining, Colorado Mining Association, Denver, Colorado

Robert Anderson, Deputy Assistant Director, Bureau of Land Management, Washington, D.C.

David Baker, Vice-President of Environmental Affairs, Newmont Mining Co., Denver, Colorado

Adele Basham, Supervisor, Water Quality Standards Branch, Nevada Environmental Protection Division, Carson City, Nevada

Douglas Bland, Director, New Mexico Mining and Minerals Division, Santa Fe, New Mexico

Steven Borell, Executive Director, Alaska Miners Association, Anchorage, Alaska

Jim Butler, Attorney, Parsons, Behle, and Latimer, National Mining Association, Salt Lake City, Utah

Cathy Carlson, Director, National Wildlife Federation, Boulder, Colorado

Nicholas Ceto, Regional Mining Coordinator, U.S. Environmental Protection Agency, Region X, Seattle, Washington

Kelly Courtright, District Mining Engineer, Bureau of Land Management, Spokane District, Spokane, Washington

Alan Coyner, Adminstrator, Nevada Division of Minerals, Carson City, Nevada

Rusty Dersch, Mining Law and Geologic Survey, U.S. Forest Service, Lakewood, Colorado

Stephen D'Esposito, President, Mineral Policy Center, Washington, D.C.

Roy Drew, Geologist, Branch of Solid Minerals, Bureau of Land Management, Lakewood, Colorado

Leo Drozdoff, Chief, Bureau of Mining Regulations and Reclamation, Nevada Division of Environmental Protection, Carson City, Nevada

James Dunn, Mine Waste Team Leader, U.S. Environmental Protection Agency, Denver, Colorado

James Edwards, Solid Minerals Group Leader, Bureau of Land Management, Lakewood, Colorado

Russell Fields, President, Nevada Mining Association, Reno, Nevada

Roger Flynn, Executive Director, Western Mining Action Project, Boulder, Colorado

Thomas Fry, Acting Director, Bureau of Land Management, Washington, D.C.

Rich Haddock, Senior Counsel, Barrick Goldstrike Mines, Inc., Salt Lake City, Utah

John Hardaway, Cripple Creek & Victor Mines, Victor, Colorado

Chris Herald, President, Crown Resources Corporation, Denver, Colorado

David Holm, Director, Colorado Division of Water Quality Control, Denver, Colorado

Hugh Ingle, President, Nevada Miners and Prospectors Association, Yerington, Nevada

Ron James, Nevada State Historic Preservation Officer, Nevada Historic Preservation Office, Carson City, Nevada

David Kiegman, Executive Director, Okanogan Highlands Alliance, Tonasket, Washington

James Kuipers, Consulting Mining Engineer, Center for Science in Public Participation, Boulder, Montana

Rodney Lentz, Area Mining Engineer, Okanogan National Forest, U.S. Forest Service, Okanogan, Washington

Tom Leshendok, Assistant Director for Minerals, Nevada State Office, Bureau of Land Management, Reno, Nevada

John Leshy, Solicitor, U.S. Department of the Interior, Washington, D.C.

Robert Loeffler, Director, Division of Mining of Water Management, Alaska Department of Natural Resources, Anchorage, Alaska

Michael Long, Co-Chair, Mine Waste Task Force, Western Governors' Association, Denver, Colorado

Mary Beth Marks, Geologist, Humboldt-Toiyabe National Forest, Elko, Nevada

Alan McKay, Associate Research Hydrogeologist, Desert Research Institute, University of Nevada, Reno

Mike Menge, Professional Staff Member, Senate Committee on Energy and Natural Resources, Washington, D.C.

Glenn Miller, Professor, Department of Environmental & Resource Sciences, University of Nevada, Reno, and Co-Chair, Mining Committee, Sierra Club, Toiyabe Chapter, Reno, Nevada

Robert Miscak, Chairman, Colorado Minerals, Energy, and Geology Advisory Board, Englewood, Colorado

Richard Mohr, Director, Environmental Operations, Phelps Dodge Corp., Phoenix, Arizona

John Mudge, Director, Environmental Affairs, Newmont Mining Company, Reno, Nevada

Tom Myers, Hydrologic Consultant, Center for Science in Public Participation, Reno, Nevada

Scott Nichols, Chief, Bureau of Minerals, Idaho Department of Lands, Boise, Idaho

Dianne Nielson, Executive Director, Utah Department of Environmental Quality, Salt Lake City, Utah

David Norman, Chief Reclamation Geologist, Division of Geology, Washington Department of Natural Resources, Olympia, Washington

Russell Plume, Hydrologist, U.S. Geological Survey, Carson City, Nevada

Geoffrey Plumlee, Chief Scientist, Central Region Mineral Resource Team, U.S. Geological Survey, Denver, Colorado

Bruce Ramsey, Assistant Director, Division of Minerals and Geology, U.S. Forest Service, Washington, D.C.

Victor Ross, Project Manager, U.S. Army Corps of Engineers, Anchorage, Alaska

Nancy Rusho, Forest Geologist, Nez Perce National Forest, U.S. Forest Service, Grangeville, Idaho

Luke Russell, Director, Environmental Affairs, Coeur d'Alene Mines, Coeur d'Alene, Idaho

Stuart Sanderson, President, Colorado Mining Association, Denver, Colorado

Michael Schwartz, Group Manager, Regulatory Affairs, Bureau of Land Management, Washington, D.C.

Christopher Sewell, Consultant, Western Shoshone Defense Project, Crescent Valley, Nevada

Laura Skaer, Executive Director, Northwest Mining Association, Spokane, Washington

Debra Struhsacker, Environmental Consultant, Reno, Nevada

Michael Turnipseed, State Engineer of Nevada, Division of Water Resources, Carson City, Nevada

Milton Ward, Chairman, President, and CEO, Cyprus Amax Minerals Company, Englewood, Colorado

Stanley Wiemeyer, Resources Contaminant Specialist, U.S. Fish and Wildlife Service, Reno, Nevada

Cy L. Wilsey, Land Manager, U.S. Homestake Mining Company, Sparks, Nevada

Shannon Work, Attorney, Givens, Funke, and Work, Coeur d'Alene, Idaho

Speakers at Public Participation Fora

Steve Balman, Glamis Imperial Corporation , Reno, Nevada

Jim Butler, Attorney, Mining Industry, Salt Lake City, Utah

Ann Carpenter, Nevada Colca Gold, Inc., Reno, Nevada

Rob Corkran, Ph.D. candidate, Colorado State University

Barbara Filas, Principal, Environmental Consulting Firm, Reno, Nevada

David Finkenbinder, National Mining Association, Washington, D.C.

Bill Goodhard, Geological Engineer, Mining Industry, Reno, Nevada

Greg Hahn, President and CEO, Summo Minerals Corporation, Reno, Nevada

Anne Hite, East Asia Gold Corp., Spokane, Washington

Harold Hoeberg, Geologist, New Mexico

Kirby Hughes, Citizens for Victor, Victor, Colorado

Ivan Irnavitz, Northwest Mining Association, Spokane, Washington

Bob Johnson, Kennecott Exploration, Reno, Nevada

Charles Knuth, Washington State Director of Gold Prospectors and Miners Association, Spokane, Washington

Mark Levin, Mining/Geological Engineer, Reno, Nevada

Frank Lewis, Small Miner, Reno, Nevada

Steve MacIntosh, Echo Bay Minerals, Republic, Washington

Donna Marie-Noel, Pyramid Lake Payute Tribe, Reno, Nevada

Ed Mangold, Citizens for Victor, Victor, Colorado

Gene McClelland, Atla Gold, Sparks, Nevada

Tim McCrum, Crowell & Moring, Washington, D.C.

Bob McQuivey, Bureau Chief, Nevada State Division of Wildlife, Reno, Nevada

Dan Mead, BHP Nevada Mining, Reno, Nevada

John Mudge, Director of Environmental Affairs, Newmont Gold Company, Reno, Nevada

Tom Myers, Center for Science in Public Participation, Reno, Nevada

Peter O'Conner, Independence Mining Company, Denver, Colorado

Ray Powers, President, Goose Creek Mining District, Spokane, Washington

Neil B. Prenn, Mine Development Associates, Mining Engineering Services, Reno, Nevada

Mike Rasmussen, Echo Bay Mining Company, Spokane, Washington

Otto Schumacher, President, Western Mine Engineer, Spokane, Washington

Laura Skaer, Executive Director, Northwest Mining Engineer, Spokane, Washington

Dave Skinner, People for the USA, Denver, Colorado

Lois Sneddon, Vice President for Conservation, Sierra Club, Reno, Nevada

Bill Upton, Placer Dome, Reno, Nevada

David Watson, Kennecott Mining Company, Salt Lake City, Utah

Anna Weiland, Spokane, Washington

Jon Winter, Environmental Specialist at Crown Jewel, Battle Mountain Gold Company, Spokane, Washington

Appendix G

Individuals Who Provided Documents to the Committee

Robert Anderson, Deputy Assistant Director, Minerals, Realty and Resource Protection, Bureau of Land Management, U.S. Department of the Interior, Washington, D.C.

Bruce Babbitt, Secretary of the Interior, Washington, D.C.

Sylvia Baca, Acting Assistant Secretary, Lands and Minerals Management, Department of the Interior, John Leshy, Solicitor, Department of the Interior, James Lyons, Under Secretary, Natural Resources and Environment, Department of Agriculture, & Charles Rawls, General Counsel, Department of Agriculture (letter to Greg Etter, Vice President and General Counsel, Battle Mountain Gold Company, March 25, 1999)

David A. Baker, Newmont Mining Corporation, Denver, Colorado

Mary Barraco, Manager of Compliance and Regulatory Affairs, Kinross Gold U.S.A., Inc., Salt Lake City, Utah

Steven G. Barringer, Singer, Brown, & Barringer, prepared for *Precious Metals Producers*, Las Vegas, Nevada

Douglas Bland, Director, New Mexico Mining and Minerals Division, Santa Fe, New Mexico

Steve Borell, Executive Director, Alaska Miners Association, Inc., Anchorage, Alaska

Gary C. Boyle, General Manager, Glamis Imperial Corporation, Winterhaven, California

Lowell Braxton, Director, State of Utah, Department of Natural Resources, Division of Oil, Gas and Mining, Salt Lake City, Utah

Neal Brecheisen, BLM Carson City Office (letter to Kinross Gold U.S.A. Inc, Carrie Eddy, March 5, 1999)

James Bright, Mining Geologist, Reno, Nevada

Michael Burnside, Northern Region U.S. Forest Service, Missoula, Montana

Jim Butler, Attorney, Parsons, Behle and Latimer, Salt Lake City, Utah

Cathy Carlson, Executive Director, National Wildlife Federation, Boulder, Colorado

Nick Ceto, Regional Mining Coordinator, U.S. Environmental Protection Agency, Region X, Seattle, Washington

Rob Corkran, Ph.D. candidate, Colorado State University

Alan Coyner, Administrator, Nevada Division of Minerals, Carson City, Nevada

Les Darling, Director, Environmental Affairs and Land Administration, Cyprus Amax Minerals Company, Englewood, Colorado

Desert Research Institute, *Technical Review and Assessment of Ground and Surface Water Relationships in the Tuscarora Mountains*, Prepared for Barrick Goldstrike Mines, Newmont Gold Company

Stephen D'Esposito, President, Mineral Policy Center, Washington, D.C.

Doug Driesner, Director of Mining Services, State of Nevada Department of Business and Industry, Division of Minerals, Reno, Nevada

Leo Drozdoff, Chief, Bureau of Mining and Regulation and Reclamation, Nevada Division of Environmental Protection, Carson City, Nevada

James Dunn, Mine Waste Team Leader, U.S. Environmental Protection Agency, Denver, Colorado

Timothy M. Dyhr, Manager, Environment, BHP Copper, Inc., Gila County, Arizona

Rich and Marilyn Fay, Citizens for Victor, Victor, Colorado

Russell Fields, President, Nevada Mining Association, Reno, Nevada

Tom Fry, Acting Director, Bureau of Land Management, Washington, D.C.

Glamis Imperial Corporation, Winterhaven, California (Chronology Handout)

Bill Goodhard, Director, Reclamation Environmental Affairs, Echo Bay, Englewood, Colorado

Robert Gosik, Kinross Gold U.S.A., Inc., Salt Lake City, Utah

Rich D. Haddock, Senior Counsel, Barrick Goldstrike Mines, Inc., Salt Lake City, Utah

Gregory Hahn, President & CEO, Summo Minerals Corporation, Arizona

Pat Hallinan, Water Quality Section, State of Washington Department of Ecology

Stanley Hamilton, Director, Idaho Department of Lands, Boise, Idaho

John Hardaway, Independence Mining Company, Inc., Victor, Colorado

Paul D. Hartley, President, ADGIS, Inc., Reno, Nevada

Christopher Herald, President, Crown Resources Corporation, Denver, Colorado

Karen Highley, Records Manager, Department of Environmental Quality, Environmental Management Bureau, State of Montana

Jane Dee Hull, Governor of Arizona, Phoenix

Bruce Humphries, Minerals Programs Supervisor, State of Colorado, Denver

Hugh Ingle, Jr., President, National Miners and Prospectors Association, Yerington, Nevada

Bob Johnson, U.S. Land Manager, Kennecott Exploration Company, Reno, Nevada

David Kimball, National Mining Association, Denver, Colorado

Chuck Knuth, Small Miner & Prospector, Gold Prospectors Association of America

James Kuipers, Consulting Mining Engineer, Center for Science in Public Participation, Boulder, Montana

Frank W. Lewis, Reno, Nevada

Ken Loda, Bureau of Land Management, Winnemucca Field Office, Winnemucca, Nevada

Bob Loeffler, Director, Division of Mining and Water Management, Department of Natural Resources, State of Alaska, Anchorage

Michael B. Long, Western Governors' Association, Denver, Colorado

Mary Beth Marks, Geologist, Humboldt-Toiyabe National Forest, U.S. Forest Service, Elko, Nevada

Timorthy McCrum, Attorney, Crowell & Moring LLP, Washington, D.C.

Daniel L. Mead, BHP Nevada Mining, Reno, Nevada

Glenn Miller, Professor, University of Nevada, Reno

Robert W. Micsak, Vice President and General Counsel, Colorado Minerals, Energy & Geology, Policy Advisory Board, Englewood, Colorado

Richard Mohr, Director, Environmental Operations, Phelps Dodge Corp., Phoenix, Arizona

Dalva Moellenberg, Attorney, Gallagher & Kennedy, P.A., Phoenix, Arizona

Tom Myer, Center for Science in Public Participation, Reno, Nevada

Newmont Gold Company, Gold Quarry Operations, Carlin, Nevada

Dianne Nielson, Executive Director, Utah Department of Environmental Quality, Salt Lake City

David Norman, Chief Reclamation Geologist, Washington State Department of Natural Resources, Spokane

Peter O'Conner, Senior Attorney, Independence Mining Company, Englewood, Colorado

Russ Plume, Hydrologist, U.S. Geological Survey, Carson City, Nevada

Geoff Plumlee, Chief Scientist, U.S. Geological Survey, Denver, Colorado

James Pompy, Manager, Reclamation Unit, Department of Conservation, State of California

Ray Powers, Moose Creek Mining District

Neil B. Prenn, Mine Development Associates, Mine Engineering Services, Reno, Nevada

James T. Price, Great Falls, Montana

Bruce Ramsey, Assistant Director, Division of Minerals and Geology, U.S. Forest Service, Washington, D.C.

Luke Russell, Director of Environmental Affairs, Coeur d'Alene Mines, Coeur d'Alene, Idaho

Stuart Sanderson, President, Colorado Mining Association, Denver, Colorado

Christopher Sewell, Consultant, Western Shoshone Defense Project, Crescent Valley, Nevada

Laura Skaer, Executive Director, Northwest Mining Association, Spokane, Washington

Lois Snedden, Vice President for Conservation, Sierra Club, Reno, Nevada

Debra W. Struhsacker, Consultant, Environmental & Government Relations, Reno, Nevada

Bob Townsend, Administrator, Minerals and Mining, State of South Dakota

Bill Upton, Placer Dome, Beowawe, Nevada

Ivan Urnovitz, Northwest Mining Association, Spokane, Washington

Milton H. Ward, Chairman and Chief Executive Officer, Cyprus Amax Minerals Company, Englewood, Colorado

David L. Watson, Director, Technical and Health, Safety, AND Environmental Quality, Kennecott Minerals Company, Salt Lake City, Utah

Kent Watson, Thompson Creek Mining Company, Clayton, Idaho

Western Governors' Association Correspondence & Comments on the 3809 Regulations, Resolution 93-006; Western Governors' Association; June 24, 1996-January 1, 1999

Francis M. Wheat, Los Angles, California

Stanley Wiemeyer, Resource Containment Specialist, U.S. Fish and Wildlife Service, Nevada Fish and Wildlife Office, Reno, Nevada

Robert D. Williams, Field Supervisor, Nevada Fish and Wildlife Office, Reno, Nevada

Cy Wisley, U.S. Land Manager, Homestake Mining Company, Sparks, Nevada

Shannon Work, Attorney, Givens, Funke & Work, Coeur d'Alene, Idaho

Appendix H

Biographical Sketches of Committee Members

PERRY R. HAGENSTEIN, *chair*, is an independent consultant on natural resources policy, economics, and management. Since 1982 he has also served as president of the Institute for Forest Analysis, Planning, and Policy, a national nonprofit research and education organization. He is the former president of Resource Issues, Inc., and former executive director of the New England Natural Resources Center. He has been a visiting professor at the Yale School of Forestry and Environmental Studies, University of Vermont, University of Massachusetts, and Vermont Law School. He served as Charles Bullard research fellow at the John F. Kennedy School of Government and Harvard Forest, Harvard University, and as a senior policy analyst for the U.S. Public Land Law Review Commission. He has been a member of the NRC Committee on Surface Mining and Reclamation, Committee on Abandoned Mine Lands (chair), Committee on Onshore Oil and Gas Leasing (chair), Committee on Earth Resources, Board on Earth Sciences and Resources, Committee on Environmental Issues in Pacific Northwest Forest Management, and Committee on Noneconomic and Economic Value of Biodiversity.

SAMUEL S. ADAMS, *vice-chair*, is an independent minerals consultant in Lincoln, New Hampshire. He is the past-president of the American Geological Institute, past-president of the Society of Economic Geologists, and former head and professor of the Department of Geology and Geological Engineering at the Colorado School of Mines. He has 24 years of industry experience as a mine and exploration geologist, exploration manager, chief geologist, and vice-president for geology and technology. He is a member of the Ad Hoc Advisory Committee to the supervisor of the White Mountain National Forest. He served as chair of the NRC Panel to Review the Mineral Resource Surveys Program Plan of the U.S. Geological Survey. He is a former member of the NRC Committee on Earth Resources and the Board on Earth Sciences and

Resources. He received his B.A. and M.A. from Dartmouth College and his Ph.D. from Harvard University.

ANNE C. BALDRIGE has worked in environmental impact evaluations and regulatory compliance in the mining industry in the United States and internationally for the last 20 years. Most recently, she was vice-president for environmental and governmental affairs, Battle Mountain Gold Company, where she helped develop a mining regulatory framework for Bolivia and submitted the first environmental impact assessment for a large operation under the new regulation. Additionally, she was responsible for the initial permitting of a mine under one of the most complex permitting scenarios in the United States, which included four federal agencies, American Indian interests, more than a dozen state and local agencies, and significant opposition to the project. She previously served as an associate with Golder Associates, Inc., principal of EIC Corporation, division head of Steffen, Robertson, and Kirsten, and reclamation specialist with the State of Colorado. Ms. Baldrige was awarded a B.S. in geology from the University of Pittsburgh and a M.B.A. in finance and accounting from Regis University.

PAUL B. BARTON, JR., recently retired from the U.S. Geological Survey. His research focuses on the genesis of mineral deposits, chemical and physical nature of ore-forming fluids, phase relations between minerals, thermodynamic properties of minerals, long-range availability of resources, and geochemistry. He was president of the Mineralogical Society of America in 1986 and president of the Society of Economic Geologists in 1979. He is a member of the National Academy of Sciences, the NRC Board on Earth Sciences and Resources, and the NRC Committee on Earth Resources.

EDWIN H. CLARK II is president of Clean Sites, Inc., in Alexandria, Virginia. He is the former secretary of natural resources and environmental control for the State of Delaware, vice-president of the Conservation Foundation, and associate assistant administrator for pesticides and toxic substances in the U.S. EPA. He holds a Ph.D. in applied economics from Princeton University. He has served as a member of the NRC Board on Environmental Studies and Toxicology.

DONALD W. GENTRY is president and chief executive officer of PolyMet Mining Corporation. From 1972 to 1998, he was on the faculty of the Colorado School of Mines, where he served as professor of mining engineering, head of the Mining Engineering Department, and dean of engineering and undergraduate studies. He has served as president of the

Society for Mining, Metallurgy, and Exploration, and president of the American Institute of Mining, Metallurgical, and Petroleum Engineers. He has been a member of the Board of Directors of Santa Fe Pacific Gold Corporation, Newmont Gold Company, and Newmont Mining Corporation. He has contributed to the restructuring of national mineral policies and related taxation issues in developing countries, including Chile, Peru, and Argentina. His expertise includes surface and underground mining and mine planning, as well as project evaluation and financing. He has published more than 100 papers. He received a B.S. from the University of Illinois, a M.S. from the University of Nevada, and a Ph.D. in mining engineering from the University of Arizona. He is a member of the National Academy of Engineering.

RAYMOND E. KRAUSS is an independent environmental planning and resource management consultant. Previously, he served as the environmental manager for the Homestake Mining Company's McLaughlin Mine. In this capacity, he managed all phases of the project from discovery and permitting through engineering, construction, operations, reclamation, and initial closure. He received the 1993 Earle A. Chiles Award for application of environmentally sound management principles to the mining and extraction of mineral resources in the intermountain West. Also, he received a Merit Award from the Soil Conservation Society for his aggressive mitigation of the environmental impact of the McLaughlin Mine. His technical skills cover the areas of pollution prevention strategies, waste rock and tailings disposal, and the prevention of acid rock drainage.

ANN S. MAEST is an aqueous geochemist whose research interests include the fate, transport, and speciation of metals and other contaminants in natural waters. She has worked as a research geochemist at the U.S. Geological Survey in Menlo Park, California; as a senior scientist at the Environmental Defense Fund in Washington, D.C.; and as a senior scientist and manager at Hagler Bailly and Hydrosphere, environmental consulting companies in Boulder, Colorado. Currently she is vice-president of Buka Environmental in Boulder. She has worked nationally and internationally on water quality and policy aspects of hardrock mining and has served on two NRC committees related to mineral resource research at federal agencies. She holds a Ph.D. in geochemistry and water resources from Princeton University, and was a NRC research associate at the U.S. Geological Survey. Her recent collaborative research efforts with fish toxicologists have focused on the effect of chemical speciation on the toxicity of metals to salmonids.

JAMES M. MCELFISH, JR., is a senior attorney and director of the Mining Center at the Environmental Law Institute, where he also works on RCRA, biological diversity, and sustainable development issues. He is lead author of two reference works on mining in the United States—*Hard Rock Mining: State Approaches to Environmental Protection* and *Environmental Regulation of Coal Mining: SMCRA's Second Decade*—and has written chapters of several treatises, including "State Environmental Law" in *The Law of Environmental Protection* and "Recreation Law" in *Sustainable Environmental Law*, both updated annually. He is a 1979 graduate of Yale Law School.

DUNCAN T. PATTEN is professor emeritus of plant biology and past director of the Center for Environmental Studies at Arizona State University. He also is an adjunct research professor with the Mountain Research Center at Montana State University. His research interests include arid and mountain ecosystems, especially the understanding of ecological processes and restoration of western riparian and wetland ecosystems. He was a senior scientist of the Bureau of Reclamation's Glen Canyon Environmental Studies, overseeing the research program to evaluate the effects of Glen Canyon Dam operations on the Colorado River riverine ecosystem. He is past-president of the Society of Wetland Scientists. He has been a member of the NRC Commission on Geosciences, Environment, and Resources; the Board on Environmental Studies; and numerous NRC committees. He chaired the National Technical Advisory Committee to the National Institute for Global Environmental Change. He received an A.B. from Amherst College, a M.S. from the University of Massachusetts at Amherst, and a Ph.D. from Duke University.

JONATHAN G. PRICE is state geologist and director of the Nevada Bureau of Mines and Geology. He is president-elect, as of July 1, 1999, of the Association of American State Geologists and chair of the Research Council of the Society for Mining, Metallurgy, and Exploration, and is past-president of the American Institute of Professional Geologists. His experience includes serving as geologist with the Anaconda Company, adjunct assistant professor at Bucknell University, geologist with U.S. Steel Corporation, and research associate, research scientist, and then program director, mineral resources, at the Bureau of Economic Geology, The University of Texas at Austin. In addition, he has served as director of the Texas Mining and Mineral Resources Research Institute, lecturer in the Department of Geological Sciences at The University of Texas at Austin, and staff director for the NRC Board on Earth Sciences and Resources. His research and publications address mineral resources, geology and geochemistry of ore deposits, igneous petrology, tectonics and geologic mapping, environmental geochemistry, and solution mining. He is a member of the NRC Committee on Earth Resources, the Board on Earth Sciences and Resources, and has served on the Panel to Review the Mineral Resource Surveys Program Plan of the U.S. Geological Survey.

RICHARD E. REAVIS is a former deputy administrator for air, water, and mining programs in the Nevada Division of Environmental Protection. In that capacity he developed policy and guidance to provide direction for state regulatory programs. Mr. Reavis served in a number of positions with the U.S. EPA, including deputy director of the Water Division for EPA Region 9, border coordinator, chief of the Arizona Branch of the Water Division, and chief of the Nevada Branch of the Water Division. He has also been awarded the EPA's Silver Medal for superior service and the U.S. Public Health Service Meritorious Service Medal. He received his B.S. degree in civil engineering from the Missouri School of Mines and Metallurgy and his M.S. degree in civil engineering from Northwestern University.

DONALD D. RUNNELLS is senior technical advisor of Shepherd Miller, Inc., a consulting environmental and geotechnical engineering firm in Fort Collins, Colorado. Previously, he served as chair of the Department of Geological Sciences at the University of Colorado, Boulder, and is now professor emeritus. His research and teaching have been in geochemistry of mineral deposits, natural waters, low-temperature geochemistry, water pollution, geochemical exploration, and geochemistry of trace substances. He has also served as a geochemist at Shell Development Company in Texas and Florida and as assistant professor of geology at the University of California, Santa Barbara. He also has been a member of the NRC Water Science and Technology Board and has participated in the Committee on Ground Water Models: Scientific and Regulatory Applications and the Panel to Review the Mineral Resource Surveys Program Plan of the U.S. Geological Survey. He received a Ph.D. and M.A. in geology from Harvard University and a B.S. in geology from the University of Utah. In 1998 he was appointed by President Clinton to the Nuclear Waste Technical Review Board.

NRC Staff

CRAIG M. SCHIFFRIES, *study director (through August 1999),* is associate executive director for special projects of the NRC Commission on Geosciences, Environment, and Resources. He previously served as director of the NRC Board on Earth Sciences and Resources. He was the first director of government affairs for the American Geological Institute, a federation of 32 national scientific and professional societies in the geosciences. He was a congressional science fellow on the staff of the U.S. Senate Judiciary Committee, where he worked on legislation to ensure that federal laws keep pace with changes in technology. As a Carnegie fellow at the Geophysical Laboratory of the Carnegie Institution of Washington, he conducted research in petrology, geochemistry, and economic geology. He holds a Ph.D. in geology from Harvard University; a B.A. in philosophy, politics, and economics from Oxford University, where he was a Marshall scholar; and a B.S. and M.S. in geology and geophysics from Yale University.

GREGORY H. SYMMES, *study director (since August, 1999)*, is associate executive director of the NRC Commission on Geosciences, Environment, and Resources. Since joining the NRC staff in 1995, he has been responsible for overseeing the review and publication of reports produced under the commission. He also directed a study for the NRC Board on Radioactive Waste Management (BRWM) that led to the report, *Peer Review in Environmental Technology Development Programs: The Department of Energy's Office of Science and Technology*, and is currently directing a BRWM study that will lead to the report, *Technologies for Environmental Management: The Department of Energy's Office of Science and Technology*. Prior to joining the NRC, he served as a research assistant professor and postdoctoral associate in the Department of Earth and Space Sciences at the State University of Stony Brook. He received his Ph.D. in geology from the Johns Hopkins University, where he studied fluid-rock interaction and mineral-fluid-melt phase equilibria in the deep crust, and his B.A. summa cum laude in geology from Amherst College

Acronyms

ACEC	Area of Critical Environmental Concern
BLM	Bureau of Land Management, an agency of the Department of the Interior
CERCLA	Comprehensive Environmental Response, Compensation, and Liability Act
CFR	Code of Federal Regulations
COSMAR	Committee on Surface Mining and Reclamation, an NRC committee that in 1979 wrote *Surface Mining of Non-Coal Minerals: A Study of Mineral Mining from the Perspective of the Surface Mining Control and Reclamation Act of 1977*
DOE	Department of Energy
EA	Environmental Assessment
EIS	Environmental Impact Statement
EPA	Environmental Protection Agency
ESA	Endangered Species Act
FLPMA	Federal Land Policy and Management Act
FONSI	Finding of No Significant Impact (under NEPA)
FS	see USFS
FWS	see USFWS
IBLA	Interior Board of Land Appeals
MMS	Minerals Management Service, an agency of the Department of the Interior
MOU	Memorandum of Understanding
NAE	National Academy of Engineering
NAS	National Academy of Sciences
NHPA	National Historical Preservation Act
NEPA	National Environmental Policy Act
NFMA	National Forest Management Act
NRC	National Research Council, the operating arm of the NAS and the NAE

NSF	National Science Foundation
RCRA	Resource Conservation and Recovery Act
SMCRA	Surface Mining Control and Reclamation Act
SPLP	Synthetic Precipitation Leaching Procedure
TCLP	Toxicity Characteristic Leaching Procedure
USFS	U. S. Forest Service, an agency of the Department of Agriculture
USFWS	U.S. Fish and Wildlife Service, an agency of the Department of the Interior
UMTRCA	Uranium Mill Tailings Radiation Control Act
USC	United States Code
USGS	U. S. Geological Survey, an agency of the Department of the Interior

Glossary

ACQUIRED LANDS Lands in federal ownership which were obtained by the government through purchase, condemnation, of gift, of by exchange. They are one category of public lands (Bureau of Land Management, 1999b).

ALLUVIUM Natural accumulations of unconsolidated clay, silt, sand, or gravel that have been transported by water, wind, or gravity to their present position.

AQUIFER A body of rock that contains sufficient saturated permeable material to conduct groundwater and to yield significant quantities of water to wells and springs.

BACKFILLING The filling in again of a place from which the rock or ore has been removed.

BACKGROUND GEOCHEMISTRY The abundance of an element in a naturally occurring material in an area where the concentration is not anomalous.

BASE METALS Those metals usually considered to be of lesser value and of greater chemical reactivity compared to the noble (or precious) metals, most commonly copper, lead, zinc and tin.

BENEFICIATION Improvement of the grade of ores by milling, flotation, sintering, gravity concentration, or other processes. Also termed "concentration".

CASUAL USE Mining activities that only negligibly disturb BLM lands and resources. Further discussion is found in Sidebar 1-3.

CLAIM The portion of mining ground held under the Federal and local laws by one claimant or association, by virtue of one location and record. Also called a "location."

CLOSURE In this report the term refers to the point at which a company permanently stops activity (although it may still retain liabilities for unforeseen environmental or safety concerns).

COMMON VARIETY MINERALS Mineral materials that do not have a special quality, quantity, character, or location that makes them of unique commercial value. On public lands such minerals are considered saleable and are disposed of by sales or by special permits to local governments. See also Sidebar 1-2.

CONCENTRATION See "beneficiation." It also refers to the amount of a material in a host (e.g., the amount of gold in a ton of ore.)

CONSTRUCTION MINERALS (OR MATERIALS) Materials used in construction, notably sand, gravel, crushed stone, dimension stone, asbestos, clay, cement, and gypsum.

COOPERATING AGENCY Any federal, state, or local agency or Indian tribe with jurisdiction by law or special expertise enabling it to cooperate with the lead agency preparing an environmental impact statement under NEPA.

CORPORATE BONDING As used in this report, the use of corporate assets as part or all of the financial assurance for the successful completion of reclamation or other corporate responsibility.

CRITICAL ENVIRONMENTAL CONCERN Describes an area under BLM management and having special attributes.

CULTURAL RESOURCES As used in this report, natural or manmade features having cultural or historical significance, such as structures, graves, religious sites, vistas, or bodies of water.

CUMULATIVE IMPACT As used in this report, the collective impacts of several operations involving human activities, including mining, grazing, farming, timbering, water diversion or discharge, and industrial processing, also includes future impacts not immediately observable.

DEVELOPMENT The preparation of a mining property so that an ore body can be analyzed and its tonnage and quality estimated. Development is an intermediate stage between exploration and mining.

DISCOVERY As used in this report, initial recognition and demonstration of the presence of valuable mineral within a claim.

DUMP A pile of ore, coal, or waste at a mine.

EMERGENCY FUNDS (re: for low-probability, post-closure events) As used in this report, funds provided to deal with unexpected failures of reclamation on closed mining sites.

EPHEMERAL STREAM A stream or reach of a stream that flows briefly only in direct response to precipitation in the immediate locality and whose channel is at all times above the water table.

EXPLORATION As used in this report, the search for valuable minerals by geological, geochemical, geophysical, or intrusive physical examination. (See also "prospecting," which in this report is considered part of exploration.)

FEDERAL LAND MANAGEMENT AGENCIES In this report the term refers to the Bureau of Land Management and the U.S. Forest Service;

management agencies not discussed here might include the National Park Service, the Department of Energy, the Department of Defense, and others.

FERROUS METALS Metals commonly occurring in alloys with iron, such as chromium, nickel, manganese, vanadium, molybdenum, cobalt, silicon, tantalum, and columbium (niobium).

FINANCIAL ASSURANCE Funding or enforceable pledges of funding used to guarantee performance of regulatory obligations in the event of default on such obligations by the permittee.

GOOD SAMARITAN ACTION An action taken for the benefit of part or all of the community at large rather than for that of the doer. In the context of this report, it usually refers to the correction of some prior detrimental environmental legacy as a convenience or as a public service, but without direct personal or institutional benefit.

GROUNDWATER Underground water.

HARDROCK Locatable minerals that are neither leasable minerals (oil, gas, coal, oil shale, phosphate, sodium, potassium, sulphur, asphalt, or gilsonite) nor saleable mineral materials (e.g. , common variety sand and gravel). Hardrock minerals include, but are not limited to, copper, lead, zinc, magnesium, nickel, tungsten, gold, silver, bentonite, barite, feldspar, fluorspar, and uranium. (BLM, 1999b) Usually refers to rock types or mining environments where the rocks are hard and strong and where blasting is needed to break them for effective mining. As used in this report, the term hardrock minerals is defined synonomous with "locatable minerals."

HEAP LEACHING As used in this report, a process for recovery of minerals from heaps of crushed ore by percolation of a solvent (such as cyanide for gold, or ferric sulfate and sulfuric acid for copper) through the heap, followed by chemical processing of the lixiviant.

LEACH PAD The surface upon which ore is piled for heap leaching, including those facilities to collect the lixivant for mineral recovery.

LEASABLE MINERALS A legal term that identifies a mineral or mineral commodity that is leasable by the federal government under the Mineral Leasing Act of 1920 and similar legislation. Leasable minerals include oil, gas, sodium, potash, phosphate, coal, and all minerals on acquired lands. See Sidebar 1-1.

LIXIVIANT A liquid medium that selectively extracts the desired metal from the ore or material to be leached rapidly and completely, and from which the desired metal can be recovered in a concentrated form.

LOCATABLE MINERALS A legal term that identifies minerals acquired through the General Mining Law of 1872, as amended. Examples are given in Table A-1. Locatable minerals are distinguished from federally owned minerals that are disposed of by leasing (see leasable minerals). In some situations, the term "hardrock minerals" is applied to locatable minerals. See Sidebar 1-1.

LOCATION See "claim." Also, the process of claiming or appropriating a parcel of mineral land.

LODE CLAIM Synonymous with "vein claim." As used in this report, a claim based on the presumption that the valuable mineral is a part of a bed-rock lode, vein, stockwork, stratum, or intrusion and is not dominantly a physical redistribution of values by surficial processes (the latter constitutes a placer deposit).

MINE An opening or excavation in the ground for the purpose of extracting minerals.

MINERAL Several other common meanings, but the following is used in this report: Any natural resource extracted from the earth for human use; e.g., ores, salts, coal, or petroleum.

MINERAL DEPOSIT A mineral occurrence of sufficient size and grade that it might, under favorable circumstances, be considered to have economic potential.

MINERAL OCCURRENCE A concentration of mineral that is considered to be valuable or that is of technical or scientific interest.

MINERAL SPECIES Term used in this report to distinguish specific mineralogical species from the unmodified term "*mineral*" as defined above.

MULTIPLE USE A combination of balanced and diverse resource uses that takes into account the long-term needs of future generations for renewable and nonrenewable resources, including, but not limited to, recreation, range, timber, minerals, watershed, wildlife and fish, and natural scenic, scientific and historical values; and harmonious and coordinated management of the various resources without permanent impairment of the productivity of the land and the quality of the environment with consideration being given to the relative values of the resources and not necessarily to the combination of uses that will give the greatest economic return or the greatest unit output. [43 U.S.C. §1702 (c)].

NOTICE-LEVEL OPERATION A mining or exploration operation on BLM land involving more than casual use but requiring that the operator submit only a Notice rather than a plan of operations. It is limited to an area of disturbance of 5 or fewer acres. See Sidebar 1-3.

OPERATIONS As used in this report, all activities and facilities involved in management, access, exploration, extraction, beneficiation, maintenance, or reclamation.

ORE The naturally occurring material from which a mineral or minerals of economic value can be extracted profitably or to satisfy social or political objectives.

OVERBURDEN Material of any nature, consolidated or unconsolidated, that overlies a deposit of useful minerals or ores.

OXIDATION As used in this report, the reaction of ores or waste with oxygen (usually above the water table); in sulfide ores this results in the release of sulfuric acid that, in the absence of neutralization, mobilizes iron, copper, zinc, and other minerals. (See also redox.)

PATENT Concerning the ownership of a mining claim: as a noun, A document that conveys title to the ground; or the process of securing a patent.

PERFORMANCE-BASED STANDARDS standards expressed in terms of a desired result or outcome rather than a method, process, or technology. See also "Technically prescriptive standards."

PHREATOPHYTE A plant that obtains its water supply from the zone of saturation or through the capillary fringe and is characterized by a deep root system.

PIT LAKE As used in this report, a lake that forms within the open pit of a mining operation.

PLACER A mineral deposit that has achieved its present distribution through the prior action of moving water or wind. Placers are usually in poorly consolidated materials and are the sources of much, but not all, tin, titanium, rare earths, diamonds, and zirconium, and some gold.

PLAN OF OPERATIONS A plan for mining exploration or development on BLM land involving more than 5 acres or a plan for mining where the operator with preexisting, valid claims intends to mine in an area of Critical Environmental Concern or a Wilderness area. See Sidebar 1-3. Also a plan required for mining or exploration on Forest Service lands whenever the Forest Service determines that the operation will result in "significant" disturbance of the land surface.

POINT SOURCE DISCHARGE Discharge of pollutant from a discernible, confined and discrete conveyance, including but not limited to any pipe, ditch, channel, tunnel, conduit, well, discrete fissure, or container.

POST-CLOSURE As used in this report, referring to the time after a property formerly used for mining has been reclaimed.

PRECIOUS METAL Any of several relatively scarce and valuable metals, such as gold, silver, and the platinum group metals.

PROSPECTING The search for outcrops or surface exposures of mineral deposits. Searching for new deposits; also preliminary explorations to test the values of lodes or placers already known to exist. (See also "exploration".)

PUBLIC DOMAIN Land owned, controlled, or heretofore disposed of by the U.S. government.

PUBLIC LAND The part of the U.S. public domain to which title is still vested in the federal government and that is subject to appropriation, sale, or disposal under the general laws.

RECLAMATION Restoration of mined land to original contour, use, or condition. But as used in this report, also describes the return of land to alternative uses that may, under certain circumstances, be different from those prior to mining.

RECORD OF DECISION Under NEPA, a concise public record that states what an agency's decision was, identifies all alternatives considered by the agency and the factors considered by the agency, and states whether all practicable means to avoid or minimize environmental harm from the alternative selected have been adopted or if not, why not.

REDOX Adjective identifying chemical reactions involving oxidation (and reduction).

RESERVED LANDS Federal lands which are dedicated to or set aside for a specific purpose or program and which are, therefore, generally not subject to disposition under the operation of all of the public land laws.

RESERVE The quantity of mineral demonstrated to be present and known to be economically producible.

SALEABLE MINERALS A legal term that defines mineral commodities that are sold by contract from the Federal Government. These are generally construction materials and aggregates. See Sidebar 1-1.

SEDIMENTARY A rock composed of sediments, or ores formed during a process of sedimentation.

SUCTION DREDGE A dredge in which the material is lifted by pumping through a suction pipe.

TAILINGS As used in this report, the waste from mineral beneficiation. They are usually regarded as liabilities, but under some circumstances they may be reprocessed to recover additional values.

TECHNICALLY PRESCRIPTIVE STANDARDS As used in this report: standards expressed in terms of the techniques to be applied. See also "Performance-based standards."

UNCOMMON VARIETY MINERALS Mineral materials that have a special quality, quantity, character, or location that makes them of unique commercial value. On public lands such minerals are locatable under the Mining Law of 1872, as amended. See Sidebar 1-2.

UNNECESSARY OR UNDUE A surface disturbance greater than what would normally result when an activity is being accomplished by a prudent operator in usual, customary, and proficient operations of similar character and taking into consideration the effects of operations on other resources and land uses, including those resources and uses outside the area of operations. Failure to initiate and complete reasonable mitigation measures, including reclamation of disturbed areas or creation of a nuisance, may constitute unnecessary or undue degradation. Failure to comply with applicable environmental protection statutes and regulations thereunder will constitute unnecessary or undue degradation. Where specific statutory authority requires the attainment of a stated level of protection or reclamation, such as in the California Desert Conservation Area, Wild and Scenic Rivers, areas designated as part of the National Wilderness System administered by the Bureau of Land Management and other such areas, that level of protection shall be met.

WASTE The part of an ore deposit that is too low grade to be of economic value at the time of mining, but which may be stored separately for possible treatment later.

WATER TABLE As used in this report, the surface separating the zone is water-saturated from the zone containing air that is freely connected to the atmosphere.

WEATHERING As used in this report, the process of decomposition of rocks or ores through the action of air and water.

WITHDRAWAL Segregation of particular lands from the operation of specified public land laws, making those laws (including the mineral location and leasing laws) inapplicable to the withdrawn lands.

YEAR EVENT The probabilistic frequency for an event of a given magnitude (e.g., a 1000-year flood).

228 AUTHORITY U.S. Forest Service regulations found at 36 CFR Part 228.

261 AUTHORITY U.S. Forest Service regulations found at 36 CFR Part 261.

3809 REGULATIONS Bureau of Land Management regulations found at 43 CFR Subpart 3809.